PAUL GUINNESS

Hodder & Stoughton

A MEMBER OF THE HODDER HEADLINE GROUP

Acknowledgements

My gratitude is due to a number of people who have helped me at various stages in the preparation of this book. First and foremost I thank Lucia Winn, friend and former colleague whose idea this publication was. She has been a constant source of encouragement and did much to ensure the success of my study visit to Brazil. I also convey my thanks to Eduardo Gradilone and Ivanese Maciel of the Brazilian embassy in London who always did their best to satisfy my requests for information.

In Brazil itself I am indebted to Professor Francisco Scarlatto of the University of São Paulo, Professor Mario Dennis of the University of Brasília and Professor Geraldo Alves de Souza of the University of the Amazon for their generous assistance during my time in Brazil.

I also want to express my gratitude to Christopher Guinness for the many hours of assistance he gave with Internet searches and with the initial preparation of many of the illustrations.

The author and publishers would like to thank the following for permission to reproduce copyright photographs in this book:

Sue Cunningham, Figures 1.10a, 1.10b, 1.26, 1.28, 3.3, 3.6, 3.9, 4.13, 5.13, 6.7, 6.16, 7.18, 8.5, 8.8, 8.10, 8.17 and 8.20; Mayor's Office Curitiba, Figures 5.20, 5.23 and 5.25; Panos, Figures 4.16 and 6.15; Popperfoto, Figure 7.8; South American Pictures, Figures 5.16, 6.10, 6.20, 6.21 and 7.13.

All other photos belong to the author.

All inside artwork by Tom Cross.

British Library Cataloguing in Publication Data
A catalogue record for this title is available from The British Library

ISBN 0 340 69732 6

First published 1998
Impression number 10 9 8 7 6 5 4 3 2 1
Year 2002 2001 2000 1999 1998

Typeset by Wearset, Boldon, Tyne and Wear.
Printed in Great Britain for Hodder & Stoughton Educational, a division of Hodder Headline Plc, 338 Euston Road, London NW1 3BH by Redwood Books

For Mary, Courtenay and Christopher

Contents

INTRODUCTION

Location and Scale

With a land area of approximately 8.5 million km², Brazil is the fifth largest country in the world (Figure 1) and by far the biggest in South America, occupying almost 48 per cent of the continent. In terms of population, Brazil also ranks fifth in the global listing with a 1995 estimate of almost 158 million people (Figure 2).

Located in the eastern central part of South America, Brazil's borders extend over 23 086 km, 7367 km of which fringe the Atlantic Ocean, stretching from the Orange River, on the border of Amapa and French Guiana, to the Chui River between Rio Grande do Sul and Uruguay. Almost all South American countries border Brazil to the north, west or south, with the exceptions of Chile and Ecuador. The result is a total of 15 719 km of international frontiers, the largest of which is with Bolivia (3126 km) and the shortest with Surinam (593 km). Only China and Russia share borders with as many countries as Brazil.

Brazil is almost as wide (4330 km) as it is long (4350 km). The Equator passes through the north of the country near Macapa leaving 93 per cent of Brazil's territory in the southern hemisphere. The Tropic of Capricorn crosses in the south near São Paulo.

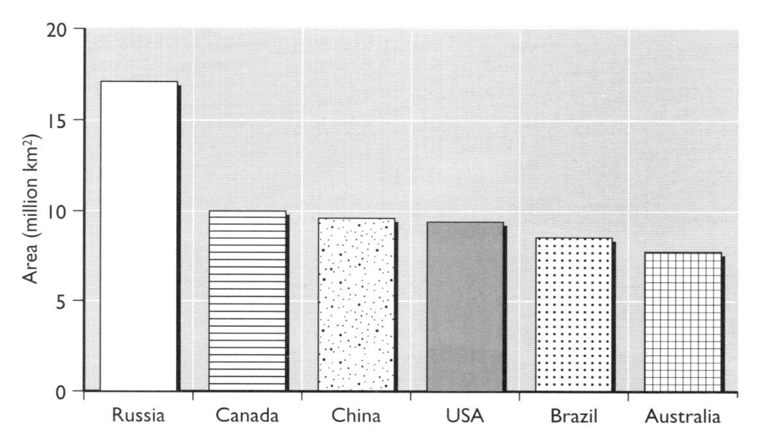

Figure 1
The largest countries in the world

Figure 2 The world's most populous nations, 1995

COUNTRY	MILLIONS
1. China	1218.8
2. India	930.6
3. USA	263.2
4. Indonesia	198.4
5. Brazil	157.8
6. Russia	147.5
The World	5702.0

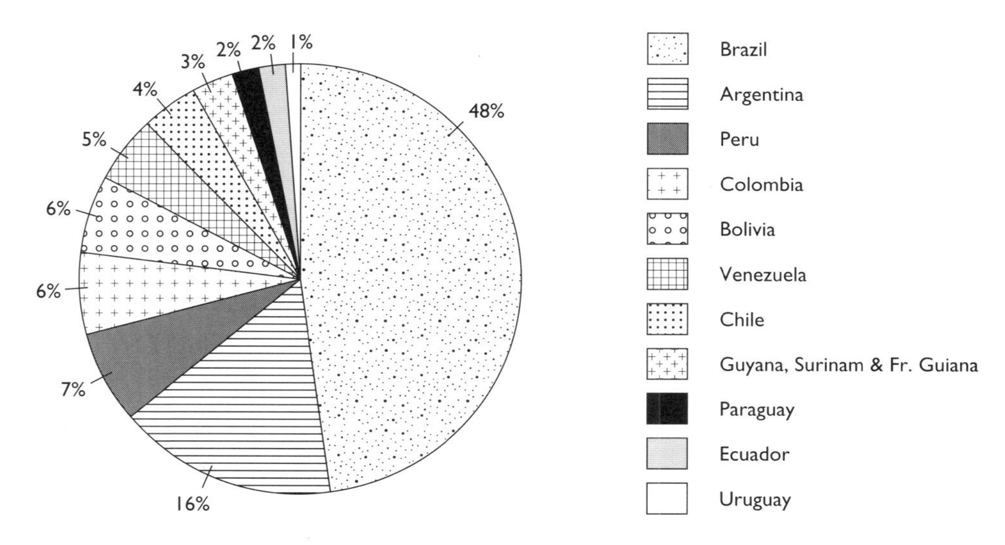

Figure 3
Brazil's share of South America in terms of area

Government and Regions

Brazil is governed by a presidential system with three independent powers: Executive, Legislative, and Judiciary. It is a Federative Republic composed of the Federal District where Brasília, the capital of the country is situated, and 26 States (Figure 4) which in turn are comprised of 4973 Municipalities. Each State has its own government, with a structure that mirrors the federal level, enjoying all the powers (defined in its own constitution) which are not specifically reserved for the Federal Government or assigned to the Municipal Councils. The States are grouped into five large regions: the North, the Northeast, the Southeast, the South, and the Centre-West. The largest region in the country, the North, occupies 45.2 per cent of the national territory.

Figure 4
The states of Brazil

Language, Race and Religion

Portuguese is the official language of Brazil. Except for the languages spoken by Indian tribes living in remote reservations, Portuguese is the only language of daily life, with no regional dialects. Brazil is the only Portuguese-speaking country in South America.

Fifty-four per cent of the population consider themselves white; 40 per cent are of mixed ethnicity; 5 per cent black; 0.5 per cent Asian; and only 0.1 per cent pure Amerindian. Although race is not generally considered to be an issue as it is in the USA for example, the white and Asian populations have on average a significantly higher standard of living than the other ethnic groups.

With the proclamation of the Republic in 1889, Brazil ceased to have an official religion although nearly 90 per cent of the population declare themselves to be Roman Catholic.

Population Growth

The first Brazilian population census, collected in 1872, announced a total of 9.9 million people (Figure 5). By the turn of the century this had increased to 17.4 million. At this time the birth rate was approximately 46 per 1000 and the death rate 30 per 1000. However, while the birth rate continued at a high level until 1960, the death rate declined steadily resulting in an increasing rate of population growth. The total population reached 50 million in the late 1940s and 100 million in the early 1970s. Since 1940 Brazil's population has nearly quadrupled.

Figure 5 Population growth, 1872–1995

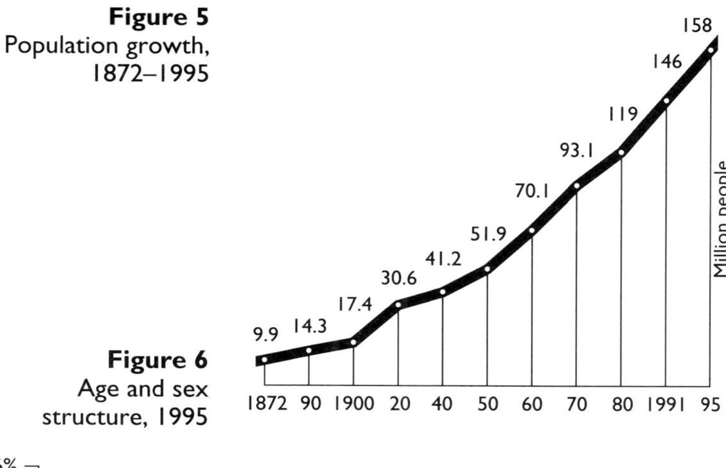

Figure 6 Age and sex structure, 1995

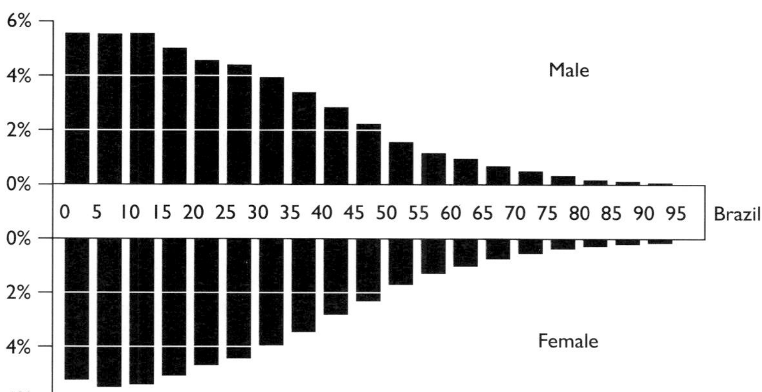

In recent decades the rate of population growth has slowed significantly. From almost 3 per cent in the 1960s, the rate dropped to 2 per cent in the 1980s, with just 1.45 per cent forecast for the 1990s. The cause has been a rapidly falling fertility rate. In the 1960s the mean fertility rate was six children per woman, in the 1970s it fell to 4.5 children and in the 1980s it dropped to under three children per woman. Such a trend is characteristic of a country experiencing significant urbanisation and industrialisation. The forecast for the year 2000 is a fertility rate of 2.2. With respect to the model of demographic transition Brazil is now in the late transition stage.

For the first time it is estimated that the population of Brazil is growing at a slower rate than the world's population as a whole. This is reflected in the youngest age group of the country's population pyramid (Figure 6) for 1995 which also shows that 33.7 per cent of Brazilians are under fifteen years of age, down from 36.4 per cent in 1985. The figure for the year 2000 is estimated at 31.8 per cent. Currently, only 7.5 per cent of the population is aged 60 years and over but this will rise to 8 per cent by the end of the decade.

Distribution and Density

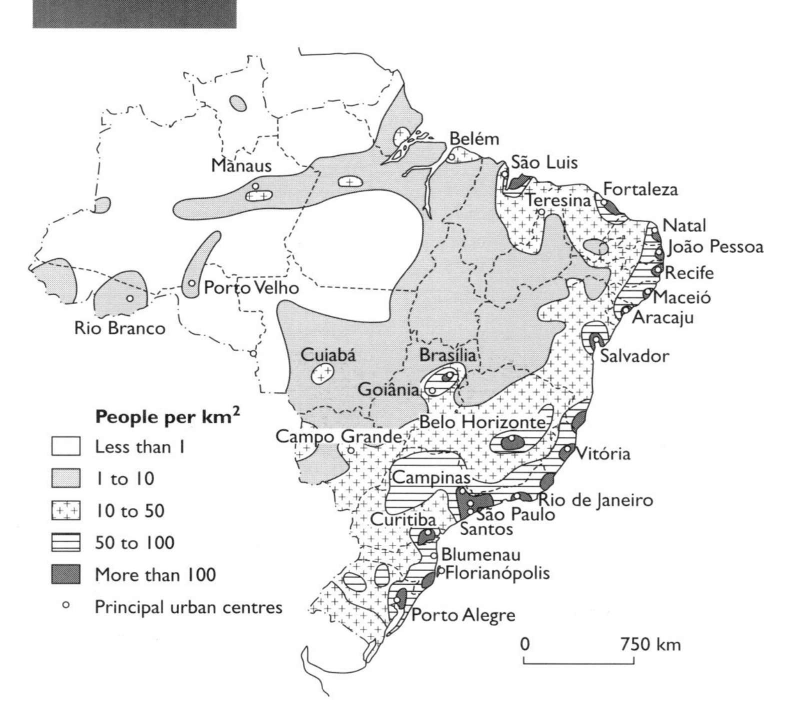

Brazil's population is very unevenly distributed (Figure 7). Highest densities are along the coast, stretching from the far north to the extreme south. In fact 80 per cent of Brazilians live within 320 km of the Atlantic coast. This pattern is largely the result of the distribution of economic activity. Density in the North is less than 3 people per km^2 making it one of the world's great areas of low density. The state of São Paulo, the hub of the national economy, has over twice as many people as the next state in demographic order (Figure 8). However it must be noted that the overall population density of Brazil is low: 19.4 people per km^2 in 1995 compared with the world average of 39. Nevertheless the impact of an increasing population density on the landscape of many areas in Brazil is clear to see. In 1970 population density was only 10.9 people per km^2 but by the year 2000 it will exceed 21.

Figure 7 Population density

Perception and Issues

On the global stage in general, Brazil is perhaps best known for its stunning achievements in football, winning the World Cup in 1958, 1962, 1970 and 1994. However in the last decade or so a range of other issues has led to a wider and more detailed interest in the country. Prominent among these are: the loss of significant areas of the Amazon rain forest; Brazil's growing economic power; the huge gap between the richest and poorest sections of society; the relatively recent return to democracy; the massive privatisation programme currently underway; and the developing attraction of the country as a tourist destination.

Figure 8 Population of the states, 1991 Census

	'000
North	**10 257**
Rondônia	1131
Acre	417
Amazonas	2103
Roraima	216
Para	5182
Amapa	289
Tocantins[a]	920
Northeast	**42 470**
Maranhão	4929
Piaui	2581
Ceará	6363
Rio Grande do Norte	2414
Paraíba	3200
Pernambuco	7123
Alagoas	2513
Sergipe	1492
Bahia	11 855
Southeast	**62 661**
Minas Gerais	15 732
Espirito Santo	2599
Rio de Janeiro	12 784
São Paulo	31 547
South	**22 117**
Paraná	8443
Santa Catarina	4538
Rio Grande do Sul	9135
Centre-West	**9412**
Mato Grosso	2023
Mato Grosso do Sulb	1779
Goias	4013
Brasília (Federal District)	1598
Total Brazil	**146 917**

[a] Tocantins was created as a new state by subdivision of the former state of Goias in 1988.
[b] Mato Grosso do Sul was created as a new state by subdivision of the former state of Mato Grosso in 1978.

Figure 9 Key dates in Brazil's history

1500	Cabral lands on the east coast of Brazil and claims it for Portugal
1549	Colonial rule established in Bahia
1695	Gold discovered in Minas Gerais
1768	Capital moved from Salvador, Bahia, to Rio de Janeiro
1789	Independence movement led by Tiradentes (who is later executed)
1815	Brazil given co-equal status with Portugal
1821	King returns to Portugal to resolve political dissension; Lisbon urges returning Brazil to colonial rule
1822	The King's son Dom Pedro declares Brazil an independent state headed by himself
1840–89	Rule of Pedro II establishes Brazil as a major Latin American power
1888	Slavery abolished
1889	Brazil becomes a republic
1920s	Movement of army officers aims to break the power of the 'coffee elite' and modernise Brazil
1930	Military coup installs Getúlio Vargas as president
1945	Army forces Vargas to resign; 'democratic elections' are held
1960	Brasília replaces Rio as capital in effort to open up the interior
1964	Military coup ousts reformist President Goulart; effective dictatorship returns to Brazil
1985	Return to civilian rule
1986	Cruzado plan launched to combat inflation; it fails
1989	Fernando Collor wins presidential election
1992	President Collor resigns on corruption charges; he is replaced by Itamar Franco
1993	Inflation exceeds 2000%
1994	Launching of *Plano Real* in July sharply reduces inflation; the architect of the plan, Fernando Henrique Cardoso, wins a landslide victory in the October presidential election

Brazil in the Global context

THE HUMAN DEVELOPMENT INDEX

The 1996 Human Development Report put Brazil in 58th place in world ranking according to the Human Development Index. The Index ranks 174 countries by a combination of life expectancy, education (as measured by adult literacy and school enrolment rates), and standard of living using Gross Domestic Product per capita. The HDI is now generally recognised as a better indicator of a country's position in the global development spectrum than the traditional measure, GDP per capita. However as Figure 1.2 shows Brazil's ranking is the same under both measures.

Brazil, along with other Latin American countries, has recorded significant progress in human development in recent decades (Figure 1.1) although a range of socio-economic problems remain to be tackled. Between 1960 and 1993 life expectancy at birth in Brazil increased from 54.7 years to 66.5 years. During the same period the infant mortality rate fell from 116 per 1000 to 57 per 1000. But while the improvement in infant mortality accelerated in the 1980s with the spread of vaccination campaigns, further improvements will depend on large-scale upgrading of urban and rural sanitation systems. While only 62 per cent of the population had access to safe water between 1975–80 this improved to 87 per cent in the period 1990–5. Similar improvements under a range of other measures were also recorded.

Figure 1.2 Human Development Index

REGION OR COUNTRY GROUP	HDI 1960	HDI 1970	HDI 1980	HDI 1993
World	0.392	0.459	0.518	0.746
Industrial countries	0.798	0.859	0.889	0.909
OECD	0.802	0.862	0.890	0.910
Eastern Europe and the CIS	0.625	0.705	0.838	0.773
Developing countries	0.260	0.347	0.428	0.563
Arab States	0.228	0.295	0.410	0.633
East Asia	0.255	0.379	0.484	0.633
Latin America and the Caribbean	0.465	0.566	0.679	0.824
South Asia	0.206	0.254	0.298	0.444
South East Asia and the Pacific	0.284	0.372	0.469	0.646
Sub-Saharan Africa	0.201	0.257	0.312	0.379
Least developed countries	0.161	0.205	0.245	0.331

Human Development Report, 1996

THE CAPABILITY POVERTY MEASURE

The capability poverty measure is an index composed of three indicators that reflect the percentage of the population with capability shortfalls in three basic dimensions of human development: living a healthy, well-nourished life; having the capability of safe and healthy reproduction; and being literate and knowledgeable. The three corresponding indicators are the percentage of children under five who are underweight; the percentage of births attended by trained health personnel; and the percentage of women aged 15 years and above who are illiterate. Out of the 101 developing countries ranked by CPM in the 1996 Human Development Report, Brazil is in thirteenth place with a CPM value of 10.0 indicating that 10 per cent of its population are capability poor, on average, in all three dimensions. Figure 1.3 shows the CPM rank and value for a selection of developing countries.

Figure 1.1
Improvements in global and regional HDI, 1960–1993

HDI RANK		LIFE EXPECTANCY AT BIRTH (YEARS) 1993	ADULT LITERACY RATE (%) 1993	COMBINED FIRST-, SECOND- AND THIRD-LEVEL GROSS ENROLMENT RATIO (%) 1993	REAL GDP PER CAPITA (PPP$) 1993	ADJUSTED REAL GDP PER CAPITA (PPP$) 1993	LIFE EXPECTANCY INDEX	EDUCATION INDEX	GDP INDEX	HUMAN DEVELOPMENT INDEX (HDI) VALUE 1993	REAL GDP PER CAPITA (PPP$) RANK MINUS HDI RANK
High human development		73.8	97.2	79	14 922	5908	–	–	–	0.901	–
1	Canada	77.5	99.0	100	20 950	5947	0.88	0.99	0.98	0.951	6
8	Iceland	78.2	99.0	82	18 640	5941	0.89	0.93	0.98	0.934	9
18	Germany	76.1	99.0	79	18 840	5941	0.85	0.92	0.98	0.920	−2
28	Malta	76.2	87.0	76	11 570	5878	0.85	0.83	0.97	0.886	6
38	Trinidad and Tobago	71.7	97.6	67	8670	5820	0.78	0.87	0.96	0.872	4
48	Mexico	71.0	89.0	65	7010	5783	0.77	0.81	0.96	0.845	−1
Medium human development		67.0	80.7	62	3044	3044	–	–	–	0.647	–
58	**Brazil**	**66.5**	**82.4**	**72**	**5500**	**5500**	**0.69**	**0.79**	**0.91**	**0.796**	**0**
68	Estonia	69.2	99.0	78	3610	3610	0.74	0.92	0.59	0.749	15
78	Tunisia	68.0	64.1	66	4950	4950	0.72	0.65	0.82	0.727	−14
88	Western Samoa	67.8	98.0	74	3000	3000	0.71	0.90	0.49	0.700	10
98	Moldova	67.6	96.4	76	2370	2370	0.71	0.90	0.38	0.663	11
108	China	68.6	80.0	57	2330	2330	0.73	0.72	0.38	0.609	3
118	Solomon Islands	70.5	62.0	46	2266	2266	0.76	0.57	0.36	0.563	−4
Low human development		56.0	48.9	46	1241	1241	–	–	–	0.396	–
128	Kenya	55.5	75.7	56	1400	1400	0.51	0.69	0.22	0.473	9
138	Laos	51.3	54.1	52	1540	1540	0.43	0.53	0.24	0.400	−2
148	Central African Republic	49.5	37.8	39	1620	1620	0.43	0.38	0.26	0.357	−15
158	Liberia	55.6	36.4	17	843	843	0.51	0.30	0.13	0.311	−3
168	Ethiopia	47.8	33.6	16	420	420	0.38	0.28	0.05	0.237	5
174	Niger	46.7	12.8	15	790	790	0.36	0.13	0.12	0.204	−17

Human Development Report 1996

Figure 1.3
Capability Poverty Measure

CPM RANK	COUNTRY	CPM VALUE
1	Chile	2.8
9	Singapore	7.7
13	Brazil	10.0
21	Venezuela	15.2
29	Malaysia	20.6
46	South Africa	30.4
63	Egypt	43.7
78	Nigeria	51.6
89	India	61.5
101	Nepal	77.3

'ECONOMIC FREEDOM'

The 1996 Index of Economic Freedom ranks 142 countries according to the range of economic restrictions (tariffs, quotas, regulations etc.), or lack of them in place. Brazil is in 94th place on the list. The Bahamas is ranked the most economically free in Latin America (12th) while Cuba has most restrictions (140th).

A NEWLY INDUSTRIALISED ECONOMY

Figure 1.4 Selected economic indicators

	AVERAGE ANNUAL GROWTH (GNP PER CAPITA) (1980–93)	GDP PER CAPITA (1993) (A) (US = 100)	MANU-FACTURING HOURLY WAGE (US$ 1992)	AVERAGE ANNUAL GROWTH IN INVESTMENT (1980–93)	AVERAGE ANNUAL GROWTH IN EXPORTS (1980–93)
OECD					
United States	1.7	100.0	11.45	2.5	5.1
Japan	3.4	84.3	18.96	5.5	4.2
Canada	1.4	81.8	12.80	3.6	5.6
Belgium	1.9	79.4	10.63[91]	3.7	4.5
France	1.6	76.8	7.88	2.1	4.5
Italy	2.1	72.1	–	1.5	4.3
Netherlands	1.7	70.0	10.44[91]	2.7	4.7
United Kingdom	2.3	69.6	10.56	4.0	4.0
Germany	2.1	68.1	14.41	2.4	4.2
First generation Asian NICs					
Hong Kong	5.4	87.1	3.28	5.0	15.8
Taiwan	–	–	5.31	–	10.0
Republic of Korea	8.2	38.9	5.25	11.8	12.3
Singapore	6.1	78.9	5.31	5.7	12.7
Second generation Asian NICs					
Thailand	6.4	25.3	0.67[91]	11.4	15.5
Indonesia	4.2	12.7	–	7.1	6.7
Malaysia	3.5	32.1	1.41[90]	6.3	12.6
Third generation Asian NICs					
Philippines	−0.6	10.8	0.48[91]	−0.1	3.4
India	3.0	4.9	0.34[89]	5.7	7.0
China	8.2	9.4	0.26	11.1	11.5
Latin American NICs					
Argentina	−0.5	33.3	–	−1.3	3.2
Brazil	0.3	21.7	1.82[88]	−0.3	5.2
Mexico	−0.5	27.5	2.11	0.1	5.4
Eastern European NICs					
Czechoslovakia	–	30.5	0.79[91]	–	–
Hungary	1.2	24.5	1.66	−1.6	2.3
Poland	0.4	20.2	1.12[91]	−1.1	2.8

(a) Based on purchasing-power parity Bank of England Quarterly Bulletin, February 1996

In Figure 1.4 Brazil, one of three newly industrialised economies recognised in Latin America, has its economic performance compared to the leading OECD (Organisation for Economic Co-operation and Development) nations and other NICs around the world. Brazil's performance during this time period was well below that of the Asian 'Tigers'.

Figure 1.5
Annual inflation 1985–95

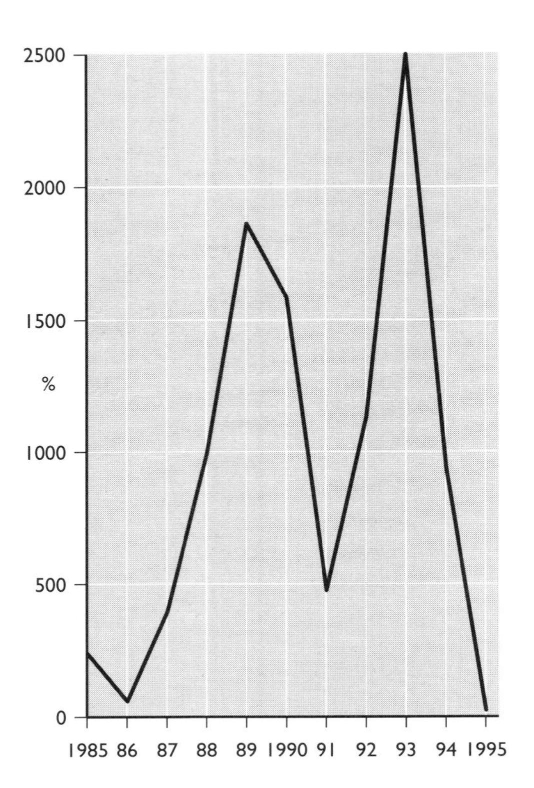

From 1993, on the initiative of then finance minister, now President Cardoso, a start was made on the preparation of an economic stabilisation programme to attempt to control once and for all the high inflation which had plagued the economy for decades (Figure 1.5). No other country in the world had experienced a longer period of chronic inflation. The culmination of this phase of economic planning was the Real Plan, introduced on 1 July 1994. At the end of June 1994, just prior to the launching of the Real Plan, inflation was running at the astonishing rate of 7000 per cent per year. For the entire year of 1995 the inflation rate was 18.8 per cent, creating much greater confidence in the economy both inside and outside Brazil. A lower rate of inflation is important not just in terms of Brazil's competitiveness abroad but also because it is a prerequisite for tackling social inequality at home. In 1995 the poorest 50 per cent of the population gained 1.2 and the richest 20 per cent lost 2.3 percentage points of their share in total income. And after the so-called 'lost decade' of the 1980s the country has returned to a rapid rate of economic growth, with a 4.2 per cent increase in GDP in 1993, 5.7 per cent in 1994 and 4.2 per cent in 1995. A recent publication from the Brazilian government states 'It should be remembered that in the 110 years between 1870 and 1980, the Brazilian economy grew faster than any other in the world.'

MERCOSUL

Mercosul (Common Market of the Southern Cone) is a customs-union established on 1 January 1995 joining Brazil, Argentina, Paraguay and Uruguay in a single market of over 200 million people. It represents 50 per cent of Latin American GDP, 43 per cent of its population, 59 per cent of its total area, and 50 per cent of its industrial production.

Figure 1.6
Evolution of exports within MERCOSUL, US$ million

	ARGENTINA	BRAZIL	PARAGUAY	URUGUAY	TOTAL
1990	1833	1320	379	712	4244
1991	1978	2309	259	679	5225
1992	2327	4098	246	689	7360
1993	3684	5394	276	1019	10 373
1994	4740	5918	377	716	11 751

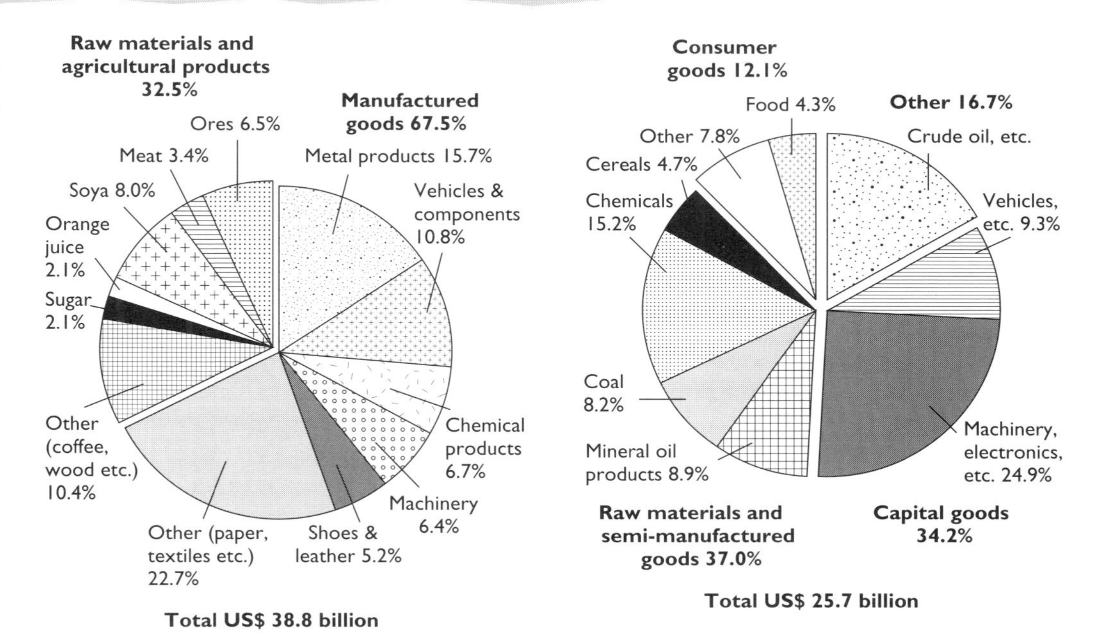

Figure 1.7 Brazil's imports and exports, 1993

In theory it is a full customs union, with a common external tariff and free trade inside it. In fact there are hundreds of national exceptions to this; but the national tariffs on them will be aligned by 2001. This is the first stage in a much wider proposed economic relationship. Mercosul is negotiating with the Andean Pact countries (Bolivia, Peru, Ecuador, Colombia and Venezuela) and Chile for the creation of a South American Free Trade Area as an important building block towards the establishment of the American Free Trade Area (AFTA) by the year 2005, as agreed during the 'Americas Summit' in Miami in December 1994. Mercosul has signed a framework agreement with the European Union to promote economic cooperation and bilateral trade.

Figure 1.8 The signing of the Mercosul agreement

Mercosul ('Mercosur' in Spanish) did not appear overnight but was the culmination of trading agreements which had grown steadily stronger in the preceding years (Figure 1.6). It began with a mid-1980s deal between Argentina and Brazil and took its present shape in 1991. Since then it has been cutting internal tariffs on an agreed schedule with the objective of creating a full customs union and, in time, free movement of labour and capital.

GLOBALISATION

The globalisation of the world economy means that the ability to compete is more important than ever before. Competition today is not so much between countries but between companies. Two of the Cardoso reforms which have had great significance in this respect are a) the privatisation programme and b) the new freedom that foreign companies now have to operate in Brazil. The latter has meant that Brazilian companies are having to modernise their practices to survive in the market.

DEBT

The Brazilian economy suffered badly in the 1980s because of its high level of borrowing. This problem is again looming large. Brazil is currently one of the most indebted nations in the world with a total foreign debt in 1996 of US $135 billion. Outside of Brazil this is a particular cause of concern to the other Mercosul countries.

? ? ? ? ? ? ? QUESTIONS ? ? ? ? ? ? ?

1. (a) Comment on the measures used in the compilation of the HDI.

(b) Select and justify two more indicators which could be incorporated into the index to give an even more comprehensive picture of contrasts in development.

2. Discuss the trends illustrated by Figure 1.1.

3. (a) What is the Capability Poverty Measure?

(b) To what extent does it help to illustrate national problems and global disparity?

4. Use an appropriate graphical method to show how trade between the Mercosul countries increased between 1990 and 1994.

5. Compare the characteristics of Brazil's exports and imports (Figure 1.7).

6. Why is a high level of foreign debt a problem to Brazil?

OVER 30 MILLION IN POVERTY

In 1995 IBGE, the Brazilian Institute of Geography and Statistics, estimated that the number of Brazilians living in absolute poverty totalled 30.4 million (Figure 1.9), 20 per cent of the Brazilian population and almost the equivalent of the entire population of Argentina. The good news was that the number of people living in absolute poverty had decreasd by 13 million since 1990. This significant decline in poverty has been attributed to the pick-up of economic activity after 1993 due to the success of the Real Plan. From 1993–5 average incomes in Brazil rose by 28 per cent, with those in lower income groups benefiting proportionately the most.

Figure 1.9 Distribution of poor in selected strata and sub-areas, 1990 and 1995

STRATA AND	1990		1995	
SUB-AREAS	NUMBER ('000S)	PROPORTION (%)	NUMBER ('000S)	PROPORTION (%)
Rural	12 227	39.26	7289	27.74
Northeast	7988	49.20	5049	32.19
South	1599	28.63	756	15.35
Mid-West	637	31.92	313	17.64
Metropolitan	12 261	28.86	9084	20.13
São Paulo (state)	3277	22.19	2640	16.79
Urban	17 483	26.85	14 065	19.16
São Paulo	1543	11.46	1539	10.24
Brazil	41 970	30.25	30 438	20.57

Gazeta Mercantil, 27 January 1997

The concept of absolute poverty applied by IBGE was not rigid as the institute took into account the level of income required to cover the basic needs of the individual in a certain time and in a specific place. In 1995 the highest poverty line was that of the São Paulo metropolitan area, while the lowest line was that of the Northeast region's rural areas. The improvement in per capita income was strenghtened by the decreasing fertility rate as families having many children are more vulnerable to poverty.

The other significant development has been the change in the location of poverty. Once predominently rural, today 23 million of those in poverty live in cities and 9 million of them in the big cities. Half of the latter live in Rio de Janeiro and São Paulo. However rural poverty is still widespread in the Northeast, where 13.5 million people classed as poor live.

THE SCALE OF INCOME INEQUALITY

Brazil has a greater disparity in income levels than most other countries in the world. In a comparison of 55 countries the distribution of income in Brazil was recorded as the worst (Figure 1.11). Economists from the Institute of Applied Economics Research in Rio de Janeiro, used as an index, the ratio between the incomes of the richest 10 per cent of the population and the poorest 40 per cent. The position has improved slightly in the last few years; in 1996 the inequality ratio in Brazil had declined to 6.36.

Figure 1.10a
Poverty in Brazil

Figure 1.10b
Affluence in Brazil

EDUCATION IS THE KEY

A recent study into the origin of such inequality concludes that the main cause is the huge variation in access to education (Figure 1.12). One of the authors of the study, Ricardo Paes de Barros, states 'There are not two Brazils. The poor and the rich live together in the same cities. They often work in the same multinational companies. The problem is that their educational background is absurdly unequal, and this results from the very poor quality of the public basic education system'. According to Barros and

PROPORTION OF INCOME EARNED BY THE	40% MOST POOR	10% MOST RICH	RATIO 10/40
Holland	22.4	21.5	1.0
Japan	21.9	22.4	1.0
Germany	20.4	24.0	1.2
Switzerland	20.1	23.7	1.2
United Kingdom	18.5	23.4	1.3
Pakistan	20.6	26.8	1.3
United States	17.2	23.3	1.4
Sweden	20.5	28.1	1.4
France	17.0	26.4	1.6
South Korea	16.9	27.5	1.6
Bangladesh	17.3	29.5	1.7
Uganda	16.6	30.7	1.8
Australia	15.4	30.5	2.0
India	16.2	33.6	2.1
Portugal	15.2	33.4	2.2
Indonesia	14.4	34.0	2.4
Argentina	14.1	35.2	2.5
Philippines	14.1	37.0	2.6
Venezuela	10.3	35.7	3.5
Colombia	11.2	44.1	3.9
Mexico	9.9	40.6	4.1
Zambia	10.8	46.4	4.3
Kenya	8.9	45.8	5.1
Botswana	7.6	42.1	5.5
Peru	7.0	42.9	6.1
BRAZIL	7.0	50.6	7.2

Figure 1.11
Inequality in the World, in increasing order, 1994

Mendonca, educational attainment explains 35–50 per cent of income inequality. Almost 15 per cent of the workforce lack a basic education and on average a college educated Brazilian earns fifteen times more than one who lacks basic education.The relationship between income and education in Brazil is stark and Figure 1.13 shows how it has been highlighted by the 1996 Human Development Report.

Figure 1.12
Education the main problem

Income study challenges myths

A recent survey conducted by Ricardo Paes de Barros and Rosane Mendonca of the Institute of Applied Economics Research has found that education is the main factor perpetuating the unequal income distribution in Brazil. Using sophisticated mathematical instruments and applied statistics drawn from data provided by the National Statistics Institute (IBGE) and other economists, the two were able to quantify the diverse factors causing Brazil's inequality. They evaluated the specific effects of race, colour, regional differences, and education on income distribution. Here are their main conclusions:

● Income inequalities in Brazil widened between 1960 and 1990, but more especially during the 1960s and 1980s. In the 1960s, uneven salary distribution affected the middle class the most, whereas in the 1980s, the poor were most affected.

● Income inequalities are concentrated in the upper layers of society. Brazil's middle class is made up of many rich, but those termed middle class are actually an upper class. The true middle class, from a statistical viewpoint, are a lower class. The richest 10 per cent of society earns three times more than the group of high wage earners composed of the 10 per cent of society right below them.

● One of the key dividing lines in Brazil's social abyss is the separation of the population between those who have a college education and those who don't.

● The regional differences—between states like São Paulo and Paraná, on one hand, and the Northeast, on the other—offer only a small explanation for the income inequalities in Brazil. "The salaries of equally productive persons in the Northeast and the Southeast are not very different," notes Barros.

● The extent of income inequality in Brazil cannot be attributed to an old and a modern sector within the economy. The poor and the rich work in the same cities, the same sectors, and the same firms.

● The segmentation of the employment market between formal and informal sectors also has little impact on the inequality of salaries.

● Men have a salary which is 42 per cent higher than women's, on average.

● The degree of differences in salary between whites and blacks in Brazil is similar to that of the U.S. In Brazil, blacks and mulattos earn 40–45 per cent less than whites.

● Differences in sex and race are not factors contributing to the high degree of inequality in Brazil.

● Differences related to how often individuals change jobs are responsible for 10 per cent of the inequality of Brazilian income.

Gazeta Mercantil, 3 March 1997

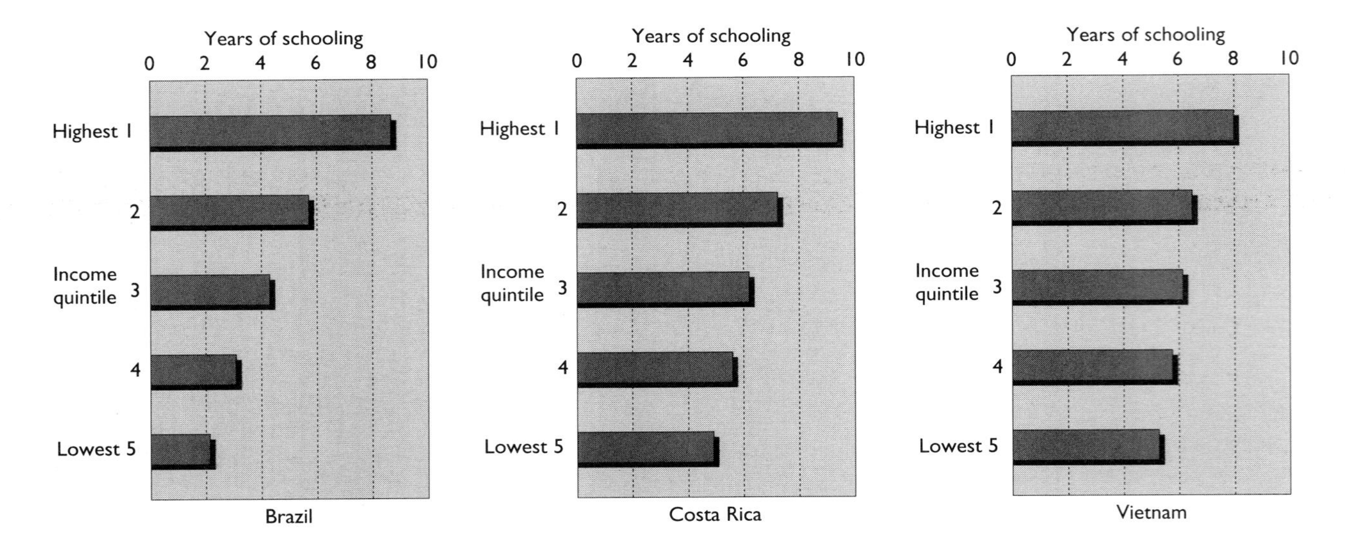

Figure 1.13 Schooling by income quintile

School is free and compulsory for the seven to fourteen age group. In the more affluent states at least 95 per cent enrol at the beginning but in the poorer states enrolment drops to 65–80 per cent. By the second year a quarter of all pupils have left, either because parents want them to work or because of dissatisfaction with the quality of education. In the country as a whole the illiteracy rate in 1991 was 16 per cent for people aged 11–14, but 20 per cent for those aged 15 and over. Both regional and urban/rural contrasts are clear (Figure 1.14). Educational standards in the best parts of the Southeast and South nearly match the first world; the poor north eastern countryside compares more with Africa.

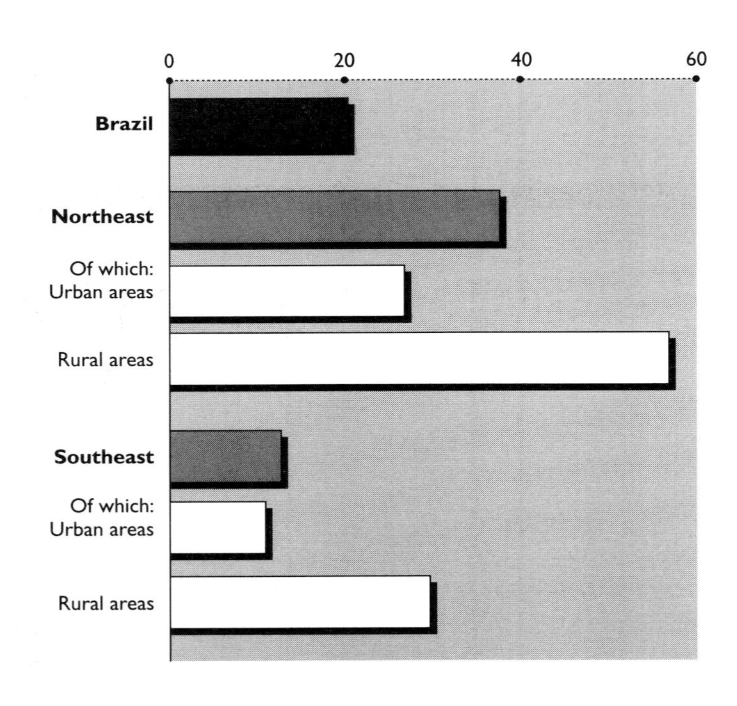

Figure 1.14 Per cent of population aged 15 and over who, in their own words, 'can't read and write a simple note'

GOVERNMENT POLICY

In the last fifty years income distribution has never been high on the national list of priorities. Economic growth, price stability and control of the foreign debt were the main goals. However the problem of income disparity at last seems to be moving towards centre stage, at least if the number of speeches on the topic by politicians and business leaders is anything to go by. President Cardoso has made education one of his main themes and already significant improvements in some areas have been recorded. Such great social inequality is a major challenge to government and society, it hinders economic development and has a clear influence on spatial patterns. Figure 1.15, a special contribution to the 1996 Human Development Report by President Cardoso, identifies the difficulty of the objective of coupling equity with economic growth and efficiency.

Humanizing growth—through equity

When we juxtapose the current debate on economic growth with the notion of human development, the first impression is that the two concepts belong to different worlds—that they don't connect. The economy reigns supreme, determining political choices and the limits of social action. And the free market emerges as a leading ideology, fostering competition and an exaggerated, narcissistic individualism that equates the realm of values with the dictates of efficiency.

Realism obliges us not to ignore efficiency. But for any development to be human, we must go beyond the logic of economics. If growth is an indispensable prerequisite, particularly in poor countries, human development will have to be sustained by values that show how economic gain acquires social meaning.

The problem is that growth based on modern technology does not always generate employment, and adopting social safety nets of a corporatist nature may jeopardize competitiveness. These difficulties are compounded by the need to reform the state, which is traditionally responsible for welfare policies and actions. Although the economic environment can change the size and management of the state, the purpose of modern governance—the well-being of citizens—must never be forsaken. Despite the criticisms and despite the weakening of social solidarity, constructing a ''state that cares for the well-being of citizens'' is a necessity. True for developed countries, this is even more true for developing countries, which are far from a welfare state.

Another significant issue is that solutions to social problems are no longer only national. Globalization limits state action and has ambivalent consequences for the development of national societies. For example, the easy transfer of capital flows can enable better resource allocation at the global level, but their volatility can provoke speculative runs on currencies, threatening the stability of entire countries.

So, we face a paradox: the demand for equity is on the increase, partly as a result of the globalization of information, yet it is directed to a state that is reducing its functions and has less control over its economic policy options.

This demand for equity—a key concept in the transition from the imperatives of economic efficiency to the realm of values—is not new. As a result of the Enlightenment, which propagated the very idea of human progress, one of the traits of Western civilization has been its permanent dissatisfaction with its social conditions.

Inherent in the ideal of progress is equity, seen as the convergence of standards of equality of opportunities—or social justice. This idea of equality has nurtured all modern utopias—from the liberal, centred on political equity, to the socialist, concerned with socio-economic equality.

The development concept has to be amplified to include the protection of human, ecological and social rights. Such complexity must be sustained by a wide participation, enabling a variety of social groups to be heard. The multiplication of non-governmental organizations, the contribution of social movements, the renewal of the meaning of political representation—all should contribute to society's redefinition of development as a way of truly humanizing growth.

In the 1960s the Third World countries searched for a new international economic order to correct the roots of international inequality, with limited success. Today, global North-South negotiations have lost force exactly when the economy is being globalized, and a homogenizing superstructure more concerned with the freedom of flows than with the reduction of inequalities is emerging.

The role of states in the international community and the way they manage multi-lateral institutions remain fundamental.

The biggest challenge for multilateral organizations is to reinvest the sense of community and to give room for international solidarity. We need a real democratization of international relations. It will not be easy, given the individualism of our time. But it is the only way to ensure that history's greatest transformations will be ethical. It is the only way development will again have a human face.

Fernando Henrique Cardoso
President of Brazil

Figure 1.15 The president speaks

? ? ? ? ? ? ? QUESTIONS ? ? ? ? ? ? ?

1. Analyse the data presented in Figure 1.9
2. (a) Examine the variations shown in Figure 1.11.

   (b) Which countries ranking higher than Brazil are you most surprised about? Explain why.
3. Comment on the findings of Barros and Mendonca (Figure 1.12).
4. (a) Assess the relationship between income and education for Brazil, Costa Rica and Vietnam shown in Figure 1.13.

   (b) Suggest reasons for the spatial variations in illiteracy illustrated in Figure 1.14.
5. Why is it perceived to be so difficult to combine the objectives of equity and efficiency (Figure 1.15)?

Regional Disparity and Development: An Overview

The first reference in national economic planning to regional disparity and development appears to be in the 1963–5 Plano Trienal. The regional problem was then addressed in more detail in a series of National Development Plans beginning in 1972. However, measures to aid individual problem regions can be traced back to earlier times. Plans to tackle the severe problems peculiar to the Northeast were formulated in the late nineteenth century and this region has been in receipt of nationally funded aid and development programmes throughout the present century. Significantly, the Constitution of 1946 set aside 4 per cent of federal reserves to development programmes for Amazonia and the São Francisco valley. Since then almost all of the country has been subject to some form of regional planning through agencies at various times (Figure 1.16).

Figure 1.16
Planning regions

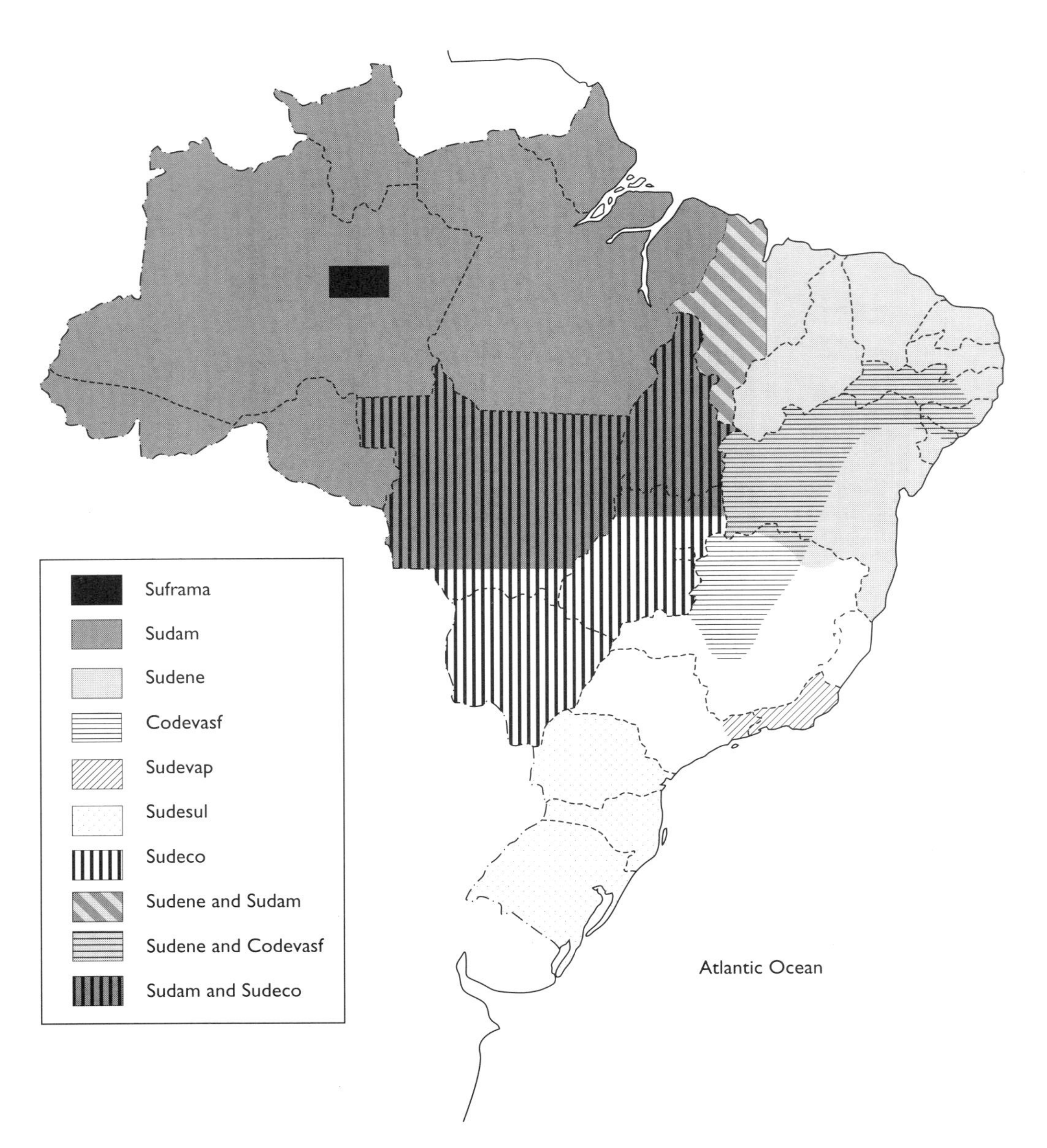

Figure 1.17 Population distribution (millions) and inhabitants per km²

	1970		1993	
REGION	POPULATION	DENSITY	POPULATION	DENSITY
North	3.6	0.9	11.0	2.9
Northeast	28.1	18.1	43.9	28.3
Southeast	39.9	43.1	64.8	70.1
South	16.5	28.7	22.7	39.5
Centre-West	5.1	3.2	9.7	6.1
Brazil	93.2	10.9	152.1	17.9

Source: IBGE

CORE-PERIPHERY THEORY AND THE SCALE OF DISPARITY

The scale of regional disparity in Brazil is considerable (Figures 1.18 and 1.19) and is characteristic of most developing countries. Although the Northeast was the first part of Brazil to be exploited, the focus of exploitation, development and investment later switched to the Southeast which was to become and remain the economic core region dominating the economy of the whole country. Core-periphery theory as proposed by Myrdal and Hirschman can be reasonably applied to the economic history of Brazil which, prior to the present century, was dominated by a series of economic cycles based on raw material exploitation (Figure 1.20). More detailed reference to individual cycles can be found elsewhere in the book.

Figure 1.18 Where the wealth is

The rich get more of it

The share of the Southeast Region in Brazil's gross domestic product (GDP) jumped from 58.18% in 1985 to 62.6% in 1995, according to research released by Confederação Nacional da Indústria (CNI), the most important institute of Brazilian industrialists. In the same period, the other four regions shrank their participation (South, −16%; Centre-West, −6%; North, −3% and Northeast, −13%).

The states of São Paulo, Rio de Janeiro, Minas Gerais and Espírito Santo make up the Southeast Region, that covers only 11% of the Brazilian territory, and gathers 44% of the country's workforce. The huge North Region, for example, holds only 4% of the work force. In the Southeast, 53.4% of the workers earn from two to ten times the minimum wage (from $220 to $1,200) a month. And 10% of this group earn more than 10 times the minimum wage a month. In the Northeast, 10.7% of the workers earn a half minimum wage or less. Only 2.9% of the region's workers earn more than 10 times the minimum wage a month.

The research shows that manufacturing has created the largest number of jobs in the Southern Region, while the Northeast depends on the creation of public jobs to absorb its workers.

Source: Gazeta Mercantil, 13 January 1997

Figure 1.19 Socio-economic indicators for Brazil and its five regions

			REGION				
INDICATOR	**YEAR**	**BRAZIL**	**SOUTHEAST**	**SOUTH**	**CENTRE-WEST**	**NORTHEAST**	**NORTH**
Rate of urbanisation	1991	75.6	88.0	74.1	81.3	80.7	59.0
% households with electricity	1996	92.0	98.0	97.0	93.0	78.0	60.0
Hospital beds per 1000 population	1992	—	4.2	4.1	4.3	3.1	2.3
% urban households served by public water supply system	1996	91.1	95.5	94.9	82.7	86.1	69.1
% illiteracy rate of children 10–14 yrs old in urban areas	1996	5.0	1.7	1.7	2.1	13.0	5.9
% children in urban areas not attending school	1996	6.3	4.7	5.3	5.2	9.5	7.9
% urban households having a telephone	1996	30.3	34.9	30.3	33.8	20.0	22.1
% urban households with TV sets	1996	91.3	95.0	93.1	88.8	83.8	84.3
% urban household units served by refuse collection	1996	87.4	92.9	95.6	89.2	72.9	64.7
Annual residential electricity consumption (KWh)	1996	2035.0	2372.0	2014.0	2145.0	1282.0	1916.0
% population under 18 yrs old	1991	41.0	37.0	38.0	42.0	47.0	48.0

The Southeast benefited from spatial flows of raw materials, capital, and labour, the latter two from abroad as well as from internal sources. The region grew rapidly through the process of cumulative causation. The impact of such cumulative causation allied to the negative backwash effect on the periphery resulted in widening regional disparity (Figure 1.21a). However, more recently some parts of the periphery, with a combination of advantages above the level of the periphery as a whole, have benefited from spread effects (trickle down) emanating from the core (Figure 1.21b). Such spread effects are clearly spatially selective and may be the result of either market forces or regional economic policy, or as is often the case, a combination of the two. The South undoubtedly has been the most important recipient of spread effects from the Southeast but parts of the Centre-West have also benefited too. In contrast the Northeast and North have found substantial economic development much more elusive.

Figure 1.20 Economic cycles

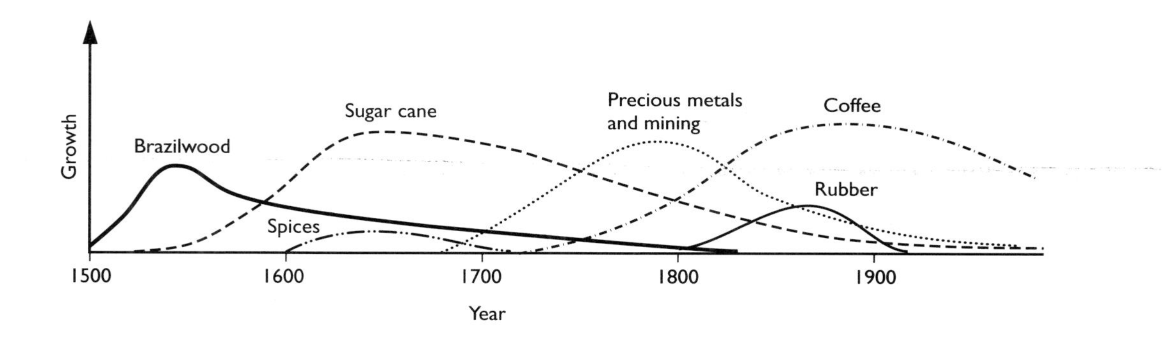

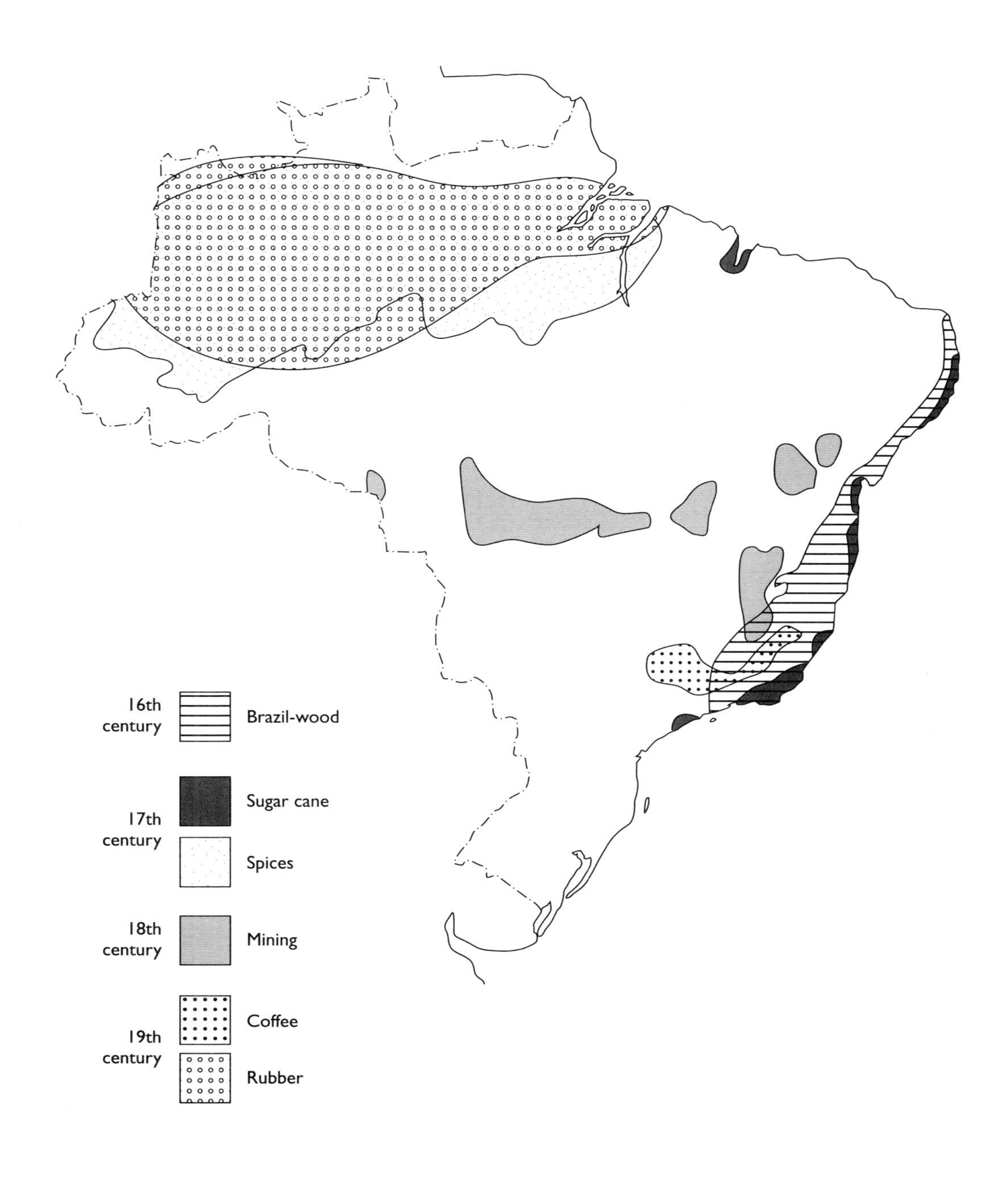

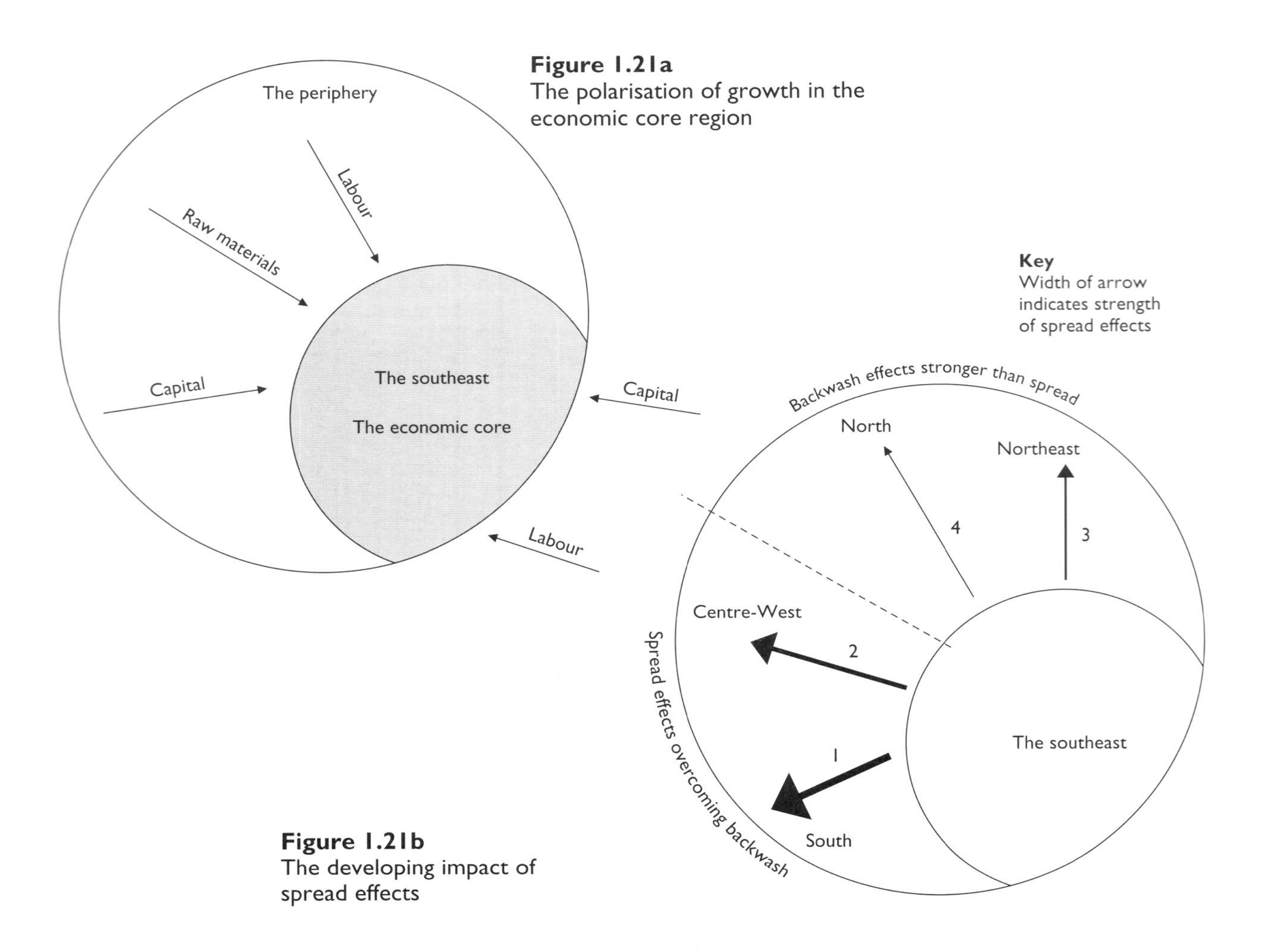

Figure 1.21a The polarisation of growth in the economic core region

Figure 1.21b The developing impact of spread effects

THE APPLICABILITY OF FRIEDMANN'S CLASSIFICATION

Thus, that part of Brazil outside the core varies greatly in its socio-economic characteristics. Here J R Friedmann's division of the periphery into upward transition areas, downward transition areas and resource frontiers seems reasonably applicable (Figure 1.22). Friedmann's model should be seen as dynamic rather than static as the economic status of a region may change over time. Initially the Southeast was a resource frontier. As money generated from its valuable raw materials was invested in industry the region went through a relatively short-lived upward transition phase before it took over from the Northeast as Brazil's economic core (Figure 1.23). In contrast the North has never been anything else but a resource frontier but it may not be that long before another label is more applicable.

Figure 1.22 Brazil's regional structure based on JR Friedmann's classification

SOUTHEAST	ECONOMIC CORE REGION	
South	Upward transition region (advanced)	Periphery
Centre-west	Upward transition region (early)	
Northeast	Downward transition region	
North	Resource frontier region	

Figure 1.23 The combination of advantages explaining the status of the Southeast as Brazil's economic core region

Climate: attractive warm temperate with adequate rainfall for agriculture, industry and settlement.

Relief and Soils: the Serra [mountain ranges] run parallel to the coast, while the interior is open, tabular upland composed of layers of lava that have been weathered to form rich terra roxa soils.

Agriculture: major world region for coffee. Also important for beef, rice, cacao, sugar cane, and fruit.

Minerals: large deposits of iron ore, manganese and bauxite. Gold is still mined.

Labour: greatest population density. Highest educational and skill levels in Brazil.

Energy: the focus of energy supply in Brazil. HEP on plateau rim; nuclear power near Rio; biomass; offshore oil and gas.

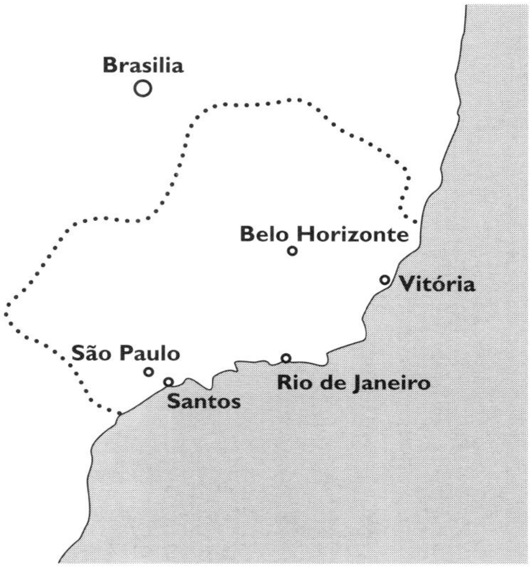

Transport: focus of road and rail networks with highest densities. Main airports and seaports. A significant pipeline network.

Government Policy: investment capital and managerial power were centralised in the Southeast, particularly in the 1950s and 60s. More recently the government has tried to spread development to other parts of the country but this has proved to be difficult.

Multinational Companies: more located in the Southeast than in the rest of Brazil. The region has grown rapidly through the process of cumulative causation.

Innovation: the centre of research and development in both the public and private sectors.

Finance: São Paulo is by far the largest financial centre in South America.

As Brazil moves along the development spectrum regional economic divergence is beginning to give way to convergence as core-periphery theory predicts.

LIMITED FUNDING

So severe were the economic problems experienced by Brazil in the 1980s and early 1990s that regional development was almost forgotten. Agencies such as SUDENE and SUDAM remained in place but the funding available was minimal. However, with the advent of the Cardoso government Brazil has reawakened to the need to tackle the disparity in development between the Southeast, South and Centre-West in relation to the North and Northeast. In addition to strengthening national unity, sustained growth in Brazil's poorest regions will also discourage migration streams, especially those directed at the saturated metropolitan areas of the Southeast.

INTRA-REGIONAL DISPARITY

It must not of course be forgotten that socio-economic inequalities exist not just between regions but also within regions. In the Southeast the agricultural areas of the sierras are some of the poorest in Brazil and the favelas of the large cities are dense foci of urban poverty. Similarly the Northeast is not totally without its wealthy areas which are mainly in and around Recife and Salvador.

THE FOUR YEAR PLAN 1996–9

The four year plan is a set of concepts, policies, strategies, guidelines and targets, which, in a coordinated way, defines the investment policy of the Brazilian government for a four year period. Its various elements involve public and private funding estimated at R$ 459 billion. The regional component of the plan is summarised in Figure 1.24.

Figure 1.24 Programmes and measures to reduce regional inequalities

Investment will be directed towards five areas of national integration and two areas of South American integration, as well as certain special areas:

1. **North-South Integration Area.** This involves the west of Bahia, the south west of Piauí, the south of Maranhão, the state of Tocantins and part of the state of Goiás, by means of the North-South Railway, the Araguaia-Tocantins Waterway and restoration works to the BR153 and the BR153/PA153.
2. **West Integration Area.** Covers Brazil's agricultural border, linking the states of Acre and Rondônia with the Centre South and South Regions, via the states of Mato Grosso and Mato Grosso do Sul. Ferronorte, the rail link between the last two states, will be the responsibility of private enterprise. The Federal Government will be responsible for the construction of the road and rail bridge over the Paraná river, thus linking Ferronorte to the rail network of São Paulo State.
3. **Northeast Integration Area.** The main projects for this region are the restoration of the São Francisco Waterway (between Pirapora in the state of Minas Gerais and Juazeiro in the state of Bahia), expansion of the Transnordestina Railway, and expansion of the Port of Suape in Recife, capital of the state of Pernambuco.
4. **South Integration Area.** This aims to improve the transport network of the region, linking it to the South East Region. The main projects are: widening of the BR 116/376/101 highways between São Paulo and Florianópolis, state of Santa Catarina; restoration of 1400 kilometres of the main federal highways in the region; construction of the Ferroeste railway between Guarapuava in the state of Paraná and Dourados in the state of Mato Grosso do Sul.
5. **Access to the Caribbean.** This involves strengthening the highway network linking Brazil to the markets of the Caribbean and the North Atlantic. The main projects are the conclusion of paving works to highway BR 174 between Manaus in the state of Amazonas and Caracarai in the state of Roraima, an extension of 624 kilometres taking the highway up to the frontier with Venezuela; and the paving of highway BR 401 between Boavista in the state of Roraima and Bonfim on the border with Guyana, which will give access to the Port of Georgetown.

continued on page 23

6. **Gateway to the Pacific.** This plan aims to link Brazil with Peru and Bolivia and give access to the ports of the Pacific. The main projects are the restoration of 350 kilometres of highway BR 317 in the state of Acre, linking the capital Rio Branco with Assis Brasil on the Peruvian frontier; restoration of 90 kilometres of the same federal highway in the state of Rondônia, linking Abuna to Guajara Mirim on the Bolivian frontier; and improvement of the conditions of navigation on the river Madeira Waterway, which links the states of Amazonas and Rondônia.

Investment will also be directed towards special projects, such as water resources for the Northeast Region, with the aim of concluding 19 irrigation projects, four dams and two pumping stations in this semi arid region of Brazil; federal government measures in the state of Rio de Janeiro, with the aim of revitalising the economy of this important state, including the establishment of the Port of Sepetiba and the Teleport of the city of Rio de Janeiro, and greater exploitation of oil and natural gas from the Campos Basin; and the necessary infrastructure for developing tourism in Brazil, through encouraging regional tourism projects such as Prodetur. This latter programme is directed towards the North East Region and includes plans for the modernisation and expansion of the airports at Natal, capital of the state of Rio Grande do Norte; Sao Luis, capital of the state of Maranhão; Fortaleza, capital of the state of Ceará; Aracaju, capital of the state of Sergipe; and Porto Seguro, a city on the south coast of the state of Bahia. Similar projects will be carried out in Amazonia and in the Pantanal of Mato Gross, directed towards ecological tourism.

Texts from Brazil, Jan/March 1997

? ? ? ? ? ? ? QUESTIONS ? ? ? ? ? ? ?

1. (a) Use an appropriate cartographic/graphical method to illustrate the data presented in Figure 1.17.

2. Study Figure 1.19:

(a) Select the five most important indicators of socio-economic disparity. Justify your choice.

(b) Rank each indicator for the five regions and then total your rankings.

(c) To what extent do your results conform to Figure 1.22?

3. Explain the processes illustrated in Figures 1.21a and b.

4. Why should Friedmann's classification be regarded as dynamic rather than static?

5. Discuss the reasons for the emergence of the Southeast as Brazil's economic core region.

6. With the aid of a good atlas use a large outline map of Brazil to show the locations of the regional projects in the four year plan 1996–9.

Regional Disparity and Development: the Northeast

THE PHYSICAL BASIS

With an area of 1.56 million km^2 the northeastern region makes up just over 18 per cent of the country's total. Most of the region's territory belongs to a vast, ancient stretch of tableland which has been smoothed by erosion. Three sub-regions can be recognised (Figure 1.25):

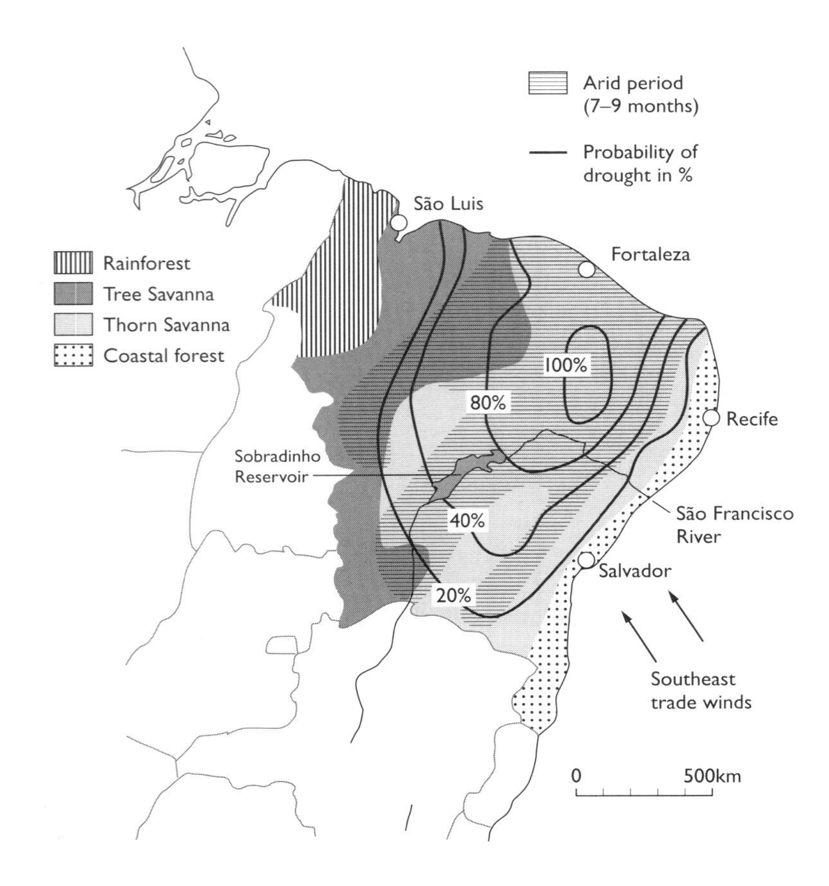

Figure 1.25
Sub-regions of the Northeast

- the 'zona da mata', the formerly wooded 200 km wide coastal strip with generally adequate precipitation and large areas of fertile red soils. The sugar economy developed in this zone and farmers also grow cacao, beans and tobacco. The native vegetation called Mata Atlantica (Atlantic Woods) has almost become extinct. The region's largest cities are situated on the coast.
- the 'agreste', a narrow transition zone between the coastal strip and the sertão. Here the most fertile lands are used for small-scale farming with subsistence crops and dairy cattle predominating.

Figure 1.26
Caatinga vegetation

- the 'sertão', the semi-arid interior consisting of thinly populated plateaus and hilly portions of the Brazilian Highlands and characterised by low and unreliable precipitation. Farmers mainly raise cattle but also grow beans, cassava, corn and cotton. Good grazing land is scarce and the soil is generally poor. The sertão is characterised by a stunted type of vegetation known as caatinga. More humid areas have woods of palm trees. That part of the sertão most susceptible to drought was designated the 'Drought Polygon' in 1951.

Figure 1.27 Living with drought

Taken by surprise, again

Brazilians count their blessings in the wetter years, instead of preparing for the dry ones

LIKE clockwork, since the early 19th century, epic droughts have punished north-east Brazil. Behind a toupee of lush coastal land lies a sprawling prairie three times the size of France and as dry as Africa's Sahel; it is struck by mild drought every three years, by a severe one every dozen. This year's drought is the worst in 40 years.

Some 10m people in more than 1,000 cities and towns covering 60% of the region have been hit. Great bridges arch over bone-dry rivers; peasants drive skinny cattle across the cracked mud of lake bottoms littered with fish bones. In Pernambuco, perhaps the driest state of all, the reservoirs have not filled since 1960. Even where there is water, it is unsafe to drink. Cholera, once eradicated from Brazil, is now an epidemic, with 4,700 cases already recorded in the north-east this year.

People in the area are not only hungry but angry. A fortnight ago rural trades-union leaders stormed the Recife headquarters of Sudene, the development agency for the north-east, and held the director and his startled officials hostage for nine hours. This helped to stir the drowsy federal government into action. Last week it released 4.7 trillion cruzeiros ($190m) in emergency-relief funds for food, water and public-works projects.

But successive years of drought have wrought permanent damage. Geographers give warning that a chunk of land nearly the size of Britain is turning into outright desert. A grim census report suggests that at least a fifth of Brazil's population, or 33m people, half of them from the north-east, are living at the ''misery'' level: chronically underpaid and underfed.

In the great droughts of the past, families would quit their scorched region for the Amazon jungle, to tap rubber latex. Or else they would migrate south to the industrial mecca of São Paulo. But now there is no exit. The Amazon rubber boom went out with Henry Ford. São Paulo, already swollen with migrants, its factories sputtering, can no longer absorb the jetsam of the ravaged backlands. Now these *flagelados* (castigated ones) drift into the poor cities of the interior, such as Petrolina and Ouricuri, where slums are burgeoning.

The region can blame man, rather than God, for its sorrows. Most of the rain, when it comes, falls in just two months of the year, March and April. If properly tapped, even the modest rains could fill enough wells and canals to rescue much of the wasted acreage. In an area threaded by rivers and streams, irrigation could redeem the semi-desert. One successful example is the Sao Francisco Valley, a sprawling oasis in the dustbowl. With huge investment and a lattice of irrigation ditches, the valley now produces a quarter of Brazil's annual fruit exports to Europe and the United States.

But, as a general rule, local and federal authorities spend the wetter years doing little more than counting their blessings. A severe drought touches off a flurry of public-works projects, which are then abandoned during the next rainy spell. As a result, every dry season is another emergency. Local authorities put out their buckets to capture the emergency-relief money. Millions of dollars allocated for drought victims are decanted through the leaky plumbing of federal, state, and local institutions, only to evaporate before they reach the people who most need assistance.

Reservoirs, cisterns, and aqueducts get built, usually by bands of unemployed and out-of-luck peasants, but they mostly end up irrigating large estates. Food packages fall into the hands of mayors and their political backers, who win votes in return for a few pounds of beans or rice. This is what north-easterners refer to as the ''drought industry''.

The Economist, 3 April 1993

Temperatures in the dry season, which lasts from May to November may rise to 42° C in the interior but can fall to 12° C in the winter. In contrast the temperature range in the coastal area is much more restricted. Southeast trade winds blowing off the Atlantic Ocean bring rain in the summer months. However, in some years the winds are weak resulting in drought in the interior (Figure 1.27). The São Francisco river provides a lifeline for much of the population in the interior, flowing 2900 km from its source before reaching the sea halfway between Salvador and Recife. There are fertile silt soils near to the river banks where farmers grow rice, maize, beans and manioc.

Due at least partly to the characteristics of the physical environment most Nordestinos, as the people of the Northeast are called, have a hard life.

Figure 1.28 Modern irrigation scene

THE PROBLEM OF DROUGHT

The region's problems were initially perceived to be the result of the periodic droughts that hit the sertão. Starvation and out-migration were the all too familiar consequences. The calamitous drought of 1877–9, after a considerable period of reasonably generous precipitation, focused attention on the region. Government reaction was to build storage reservoirs to reduce the detrimental impact of the dry season and drought years. However this did nothing to rectify the other causes of regional distress – historical legacies, an extremely high dependence on agriculture, a negative industry mix, social deprivation and the general thrust of national development which favoured the Southeast.

In the late 1940s attempts were made to implement an integrated river basin development scheme in the São Francisco valley, the region's only major perennial river. This was to involve irrigation, navigation, and flood control, but the end product fell far short of the initial plans. However, the construction of a large power plant at the Paulo Afonso Falls during the same period did reduce the region's massive power shortfall and provided some impetus to development.

A new strategy, based on the introduction of drought resistant crops, was adopted following the drought of 1951. A development bank was established to provide the credit necessary to sustain farmers over the time required for the new crops to mature. This initiative also had limited success and, following the 1958 drought, a more comprehensive approach to regional development was adopted.

SUDENE

In 1959 SUDENE (Superintendencia do Desenvolvimento do Nordeste) was established.

Figure 1.29 GDP and GDP per capita in constant prices for 1993

	GDP (US$ BILLIONS)		GDP "PER CAPITA" (US$)	
YEAR	**BRAZIL**	**NORTHEAST**	**BRAZIL**	**NORTHEAST**
1960	100.9	14.7	1458.0	663.0
1970	182.1	20.8	1962.0	740.0
1980	416.4	48.1	3510.0	1380.0
1990	482.4	66.3	3360.0	1592.0
1993	507.2	65.3	3343.0	1486.0

Its objectives were to reduce the socio-economic gap between the Northeast and the rest of the country (Figure 1.29), coordinate federal investments in the region, and improve resistance to drought. The emphasis in SUDENE plans was initially on infrastructural projects and two-thirds of its first budget was allocated to roads and power. In its second phase of planning, industrial development became a priority in order to reduce dependence on primary activities and to absorb labour. Substantial investment in industry occurred resulting in a welcome diversification of the region's industrial structure. Not surprisingly, investment was spatially concentrated to the benefit of the main urban areas, in particular Recife, Salvador and Fortaleza. Seventy per cent of the industrial projects were located in only 11 urban areas, which in 1970 had less than one-fifth of the region's population. In addition the capital intensive nature of many of the incoming firms meant that the labour absorption effect was extremely limited and in 1969 the agency acknowledged that the industrialisation policy had failed in this respect. Modernisation of existing traditional industries which became necessary for them to remain competitive reduced their labour forces. Industrial investment in the sertão, the most deprived zone in the Northeast was later in date, smaller in scale and generally less dynamic in character.

In another attempt to grapple with the problems of low incomes and excess labour, the focus switched to the perceived underpopulation of the Amazon region. The notion was that the problems of both regions could be solved at once. More recently SUDENE has moved to a more rounded strategy involving both cottage and large scale industry, agricultural assistance and the promotion of subsistance as well as commercial crops to avoid some of the worst affects of drought.

Figure 1.30 The Northeast, indicators of development

INDICATOR	1960	DATA FOR MOST RECENT YEAR	
GDP agriculture (%)	30.5	8.3	1993
GDP Industry (%)	22.1	29.6	1993
GDP Services (%)	47.4	62.1	1993
Paved roads (km)	8000	40 300	1993
Airports	9	15	1994
Electricity generating capacity (MW)	200	7800	1994
Education-matriculation third grade	14 104	247 175	1991
Income per capita NE/Brazil (%)	41.9	44.4	1993
Adult literacy (%)	34	64	1988
Life expectancy (years)	41	59.8	1991
Infant mortality (per 1000)	166	80	1992

NORTHEASTERN PACT 1996

The most recent development plan for the Northeast, the 'Pacto Nordeste', was published in January 1996. Coordinated by SUDENE, a number of strategies have been designed to enable the Northeast to further its development. Major elements of the plan include:

- consolidation of the economic basis of the region through investment in steel production, oil refining, car assembly and other manufacturing industries;
- modernisation of agriculture;
- utilisation of reservoirs for the development of aquaculture;
- expansion of the tourist industry;
- improvement of infrastructure in terms of irrigation, energy supply, transport and communications;
- training and technical development in schools, colleges and in the workplace;
- intensifying and extending health programmes covering risk groups. Improvements to water supply and sewerage will be an important element of general health improvement.

Priority has been given to water in terms of: irrigation, building new dams and reservoirs, linking river systems by canals, and improving the supply of drinking water. An important player in the development process is FINOR (Northeast Investment Fund) which is operated by the Banco do Nordeste do Brasil under the supervision of SUDENE.

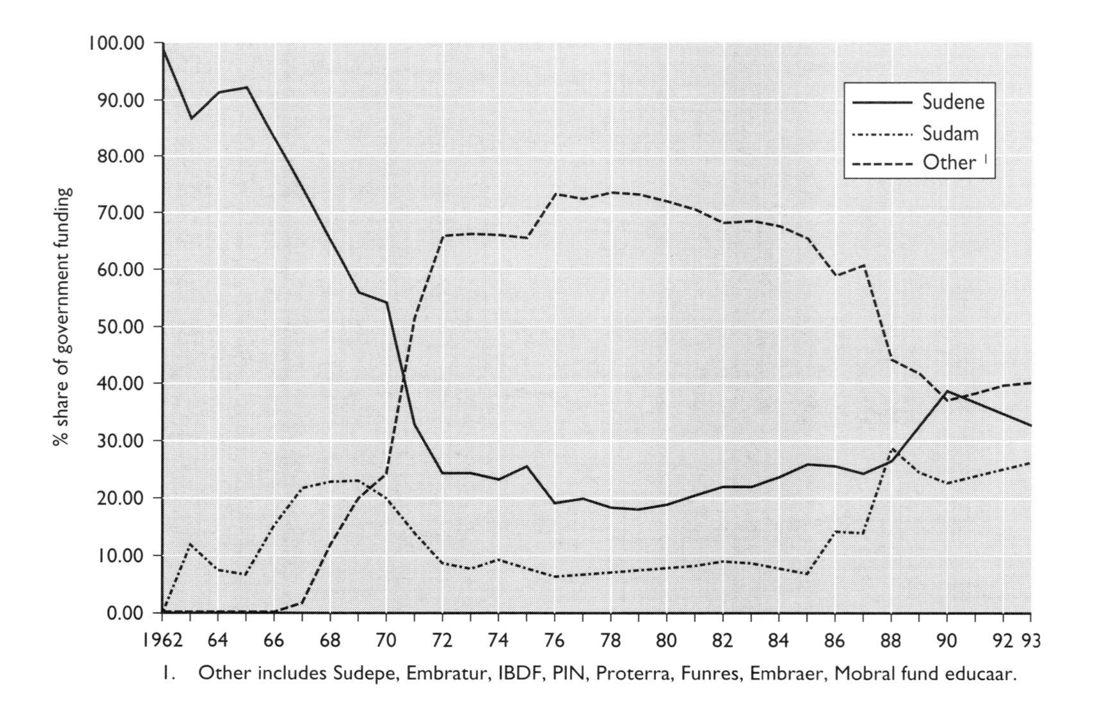

Figure 1.31 The position of SUDENE and SUDAM in terms of government funding for incentive schemes, 1962–93

NORTHEAST FRUIT GROWING DEVELOPMENT PROGRAMME

A new large-scale irrigation project (part of the Northeastern Pact) aims to turn the Northeast into 'a virtual California' according to Agriculture Ministry executive secretary Ailton Barcelos. The objective is to become the world's largest fruit exporter using Chile's distribution network. Under the programme, one million hectares in the São Francisco and Parnaíba River basins, along with other semi-arid areas in the region, will be irrigated. In comparison, Chile, one of the leading fruit export countries has 250 000 hectares under irrigation.

Sales of temperate climate fruits on the world market totalled $17.4 billion in 1996 compared to only $1.6 billion for tropical fruits. The view in Brazil is that there is no reason why sales of tropical fruits should not match temperate varieties in ten years time. An added incentive is that Chile's distribution network is partly idle between October and March, when a major portion of tropical fruits are harvested. The programme, which could create one million direct jobs in the region, is seeking $900 million a year in government funding.

RECENT INDUSTRIAL INVESTMENT

The Northeast has experienced somewhat of a resurgence in the last few years due to a combination of greater national economic stability and the objective of central government to direct a greater share of investment into the region. Price stability and a healthy rate of economic growth have encouraged both domestic and foreign investment. The relatively low wage rates in the region and the tax breaks offered by government are now proving an attractive combination (Figures 1.32 and 1.33). State governments have also been active in employing modern marketing strategies to entice new firms. The competition between the states in the Northeast to attract inward investment has been intense at times.

Figure 1.32 The factory makes the town

Factory earns more than city

At first sight Sobral, a small town 137 miles west of Fortaleza, where women still do their laundry in the unpolluted waters of the Acaraú River, does not look like a place in the throes of industrialization. The small clay houses, a few clandestine buses and motorbike-taxis contrast with the modern footwear plants built there by Grendene, based in the southern state of Rio Grande do Sul.

The bonus of cheap labor landed Grendene in Sobral a few years ago. According to the company, an employee in Ceará costs Grendene R$ 387 a month on average, or 23% less than one in its home state.

Grendene already runs two production units in Sobral and is now building a third, expected to be ready in late June and create 2,000 jobs. The company's 4,800-strong payroll costs it R$ 1.5 million a month, 25% more than Sobral's municipal revenues of R$ 1.2 million. Last year, sales stood at R$ 293.6 million, while the town's budget for 1997 has been estimated at R$ 48 million.

Gazeta Mercantil, 24 February 1997

Figure 1.33 Lowering the cost of beer

State will brew its own beer

Beer prices in Ceará are expected to drop by 20%, when the Antarctica and Kaiser breweries begin operations in the state, and local brews replace the current supplies from plants in Bahia and Rio de Janeiro, bearing the load of transportation costs in their prices.

Following the recent wave of other industries – like apparel, footwear, food, and textiles – moving to the northeast on account of lower labor costs and tax breaks, Antarctica and Kaiser are building their first plants in Ceará. Antarctica is expected to spend $160 million and Kaiser another $100 million. Kaiser production in Pacatuba, 20 miles south of Fortaleza, will begin next year, and the Antarctica plant, in Aquiraz, 20 miles southeast of Fortaleza, should open up in mid-1999.

Gazeta Mercantil, 24 February 1997

In the five years preceding 1997 the state of Ceara attracted 378 new production facilities with a combined investment of $5.2 billion. The surge began with Brazilian companies with headquarters in the Southeast and South but foreign companies are becoming increasingly represented. To start operations in Ceara, companies are granted a 45 per cent Value Added Tax (ICMS) break in the form of a loan, if located in the Fortaleza metropolitan region, or a 75 per cent ICMS break if outside this area. The loans can be repaid after a three year period. Terms for the renewable benefit can be extended to thirteen years.

The state of Maranhao has recently mounted a major exercise to attract companies from Taiwan, expecting to benefit from $1 billion of Taiwanese investments by the year 2000. Taiwanese companies list Maranhão's locational advantages as cheap labour, an abundance of raw materials and its geographical location (a shortcut to US markets). An example is the proposed $20 million Ensure cast-alloy rim plant located next to the Alumar complex which will directly supply it with hot aluminium. The end product will be exported through the nearby Itaqui deep-sea port to the US. At present Ensure imports aluminium from Australia into Taiwan with its finished product taking longer and incurring greater costs on the export route to the US.

? ? ? ? ? ? ? QUESTIONS ? ? ? ? ? ? ?

1. Compare GDP per capita in the Northeast to Brazil as a whole between 1960 and 1993.
2. Comment on the socio-economic improvements illustrated by Figure 1.30.
3. What are the physical reasons for the drought problem in the region?
4. Describe the change in SUDENE's share of government funding for incentive schemes between 1962 and 1993.
5. Outline the main elements of the 1996 Northeastern Pact.
6. What are the reasons for the recent increase in industrial investment in the Northeast?

A look at the different regions

The physical resource base in Brazil is extremely broad, encompassing a wide range of flora and fauna, a considerable climatic range, and a large geomorphological diversity. The major ecosystems (Figure 2.1) are:

- the Equatorial rain forest of the Amazon (the Selvas);
- the Cerrado;
- the Atlantic tropical forest;
- the Caatinga;
- the Pantanal;
- the coastal ecosystems.

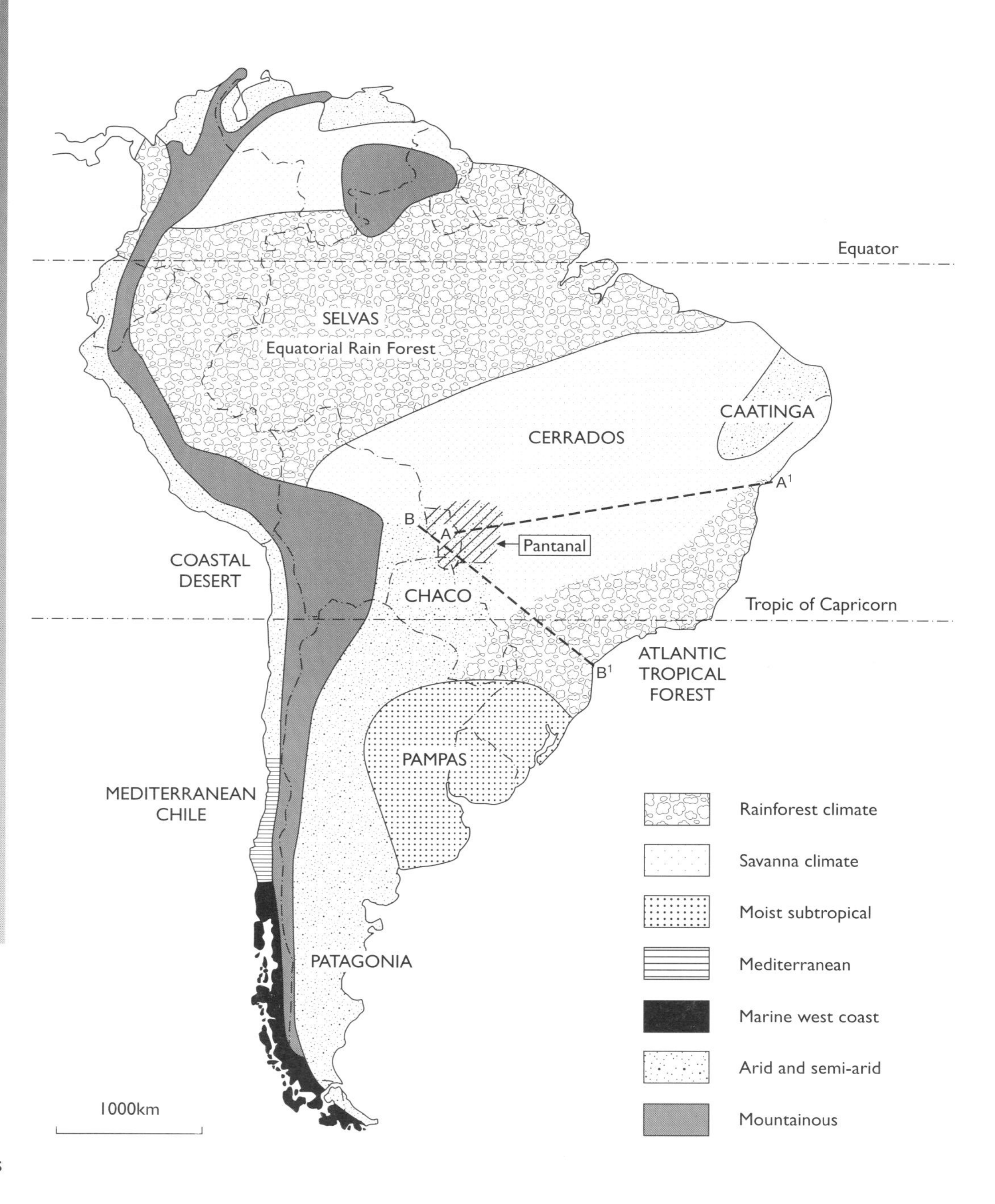

Figure 2.1
Natural regions

Figure 2.2 The Cerrado between Brasília and Goiânia

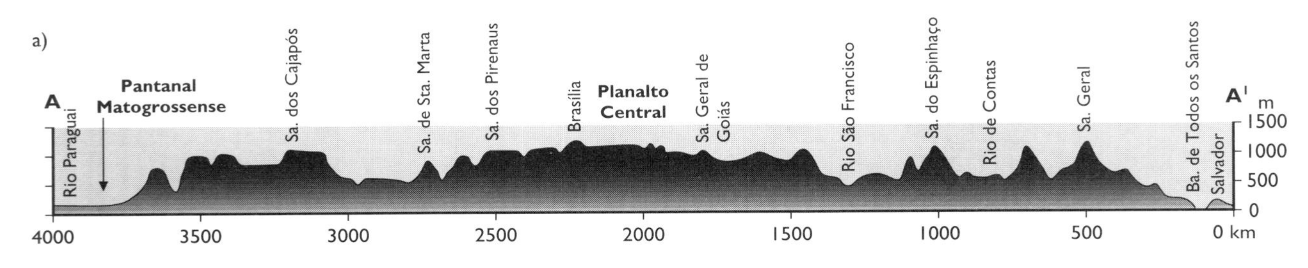

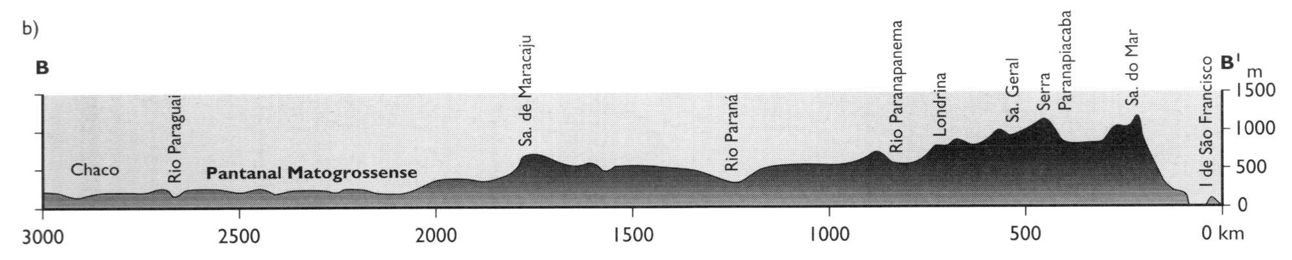

Figure 2.3a and b Cross sections from the interior to the coast

The Cerrado (Figure 2.2) is a huge, contiguous area in the central region of the country, covered with savannah-like vegetation, combining sparse scrub trees with drought-resistant grasses. It experiences a long dry season from March to August/September when humidity levels can be lower than those in the Sahara desert. Soils are generally acidic with low fertility due to large aluminium concentrations. The Cerrado occupies the Planalto Central illustrated in Figure 2.3a.

The Atlantic tropical forest covered as much as 624 000 km^2 at the time of European discovery. The forest formed a contiguous strip along the coast. Today only 4 per cent of the primary forest remains, mostly in the states of Rio de Janeiro and São Paulo. Although seriously depleted, the Atlantic Forest still shelters a valuable genetic bank. A number of mountain chains – Serra da Mantiqueira, Serra Geral, Serra do Mar – follow the south eastern coastline. In general they are covered with a beautiful secondary forest (Figure 2.4) with some rare patches of primary forest. The steepness of the Serra do Mar is illustrated by Figure 2.3b.

Figure 2.4
The Atlantic tropical forest along the Serra do Mar

A significant area of the Northeast suffers from serious drought problems. The climate is semi-arid with 500–800 mm annual rainfall, but with great variations and occasional massive droughts. The vegetation is well adapted to the conditions; much of it is a dry deciduous thorn scrub, with tree species that have special drought tolerance. This is the 'caatinga' or white forest, named for its appearance in the dry season.

Brazil's coastline presents at least four major ecosystems:

1. the equatorial coast characterised by marshes and river or ocean-flooded woods;
2. the northeastern coast, with its marshes, 'restingas' (very thin strips of land enclosing salt-water lagoons), sea reefs and granite barriers;
3. the southeastern coast, marked by rocky formations, projecting cliffs, marshes, restingas and lagoons, and
4. the southern coast, sheltering freshwater marshes with restingas and lagoons.

The rain forest of the Amazon, and the Pantanal are examined in detail in the following sections.

The Rain Forest of the Amazon

ECOSYSTEM CHARACTERISTICS

The Amazon is the world's largest rain forest and the major ecosystem in Brazil in terms of both geographical area and biodiversity. Its land area represents 40 per cent of Brazil's territory. One thousand eight hundred species of birds and 250 varieties of mammals are sheltered by an infinite assortment of trees and plants. It is one of the few places on earth where new species of plants and animals remain to be discovered. The Amazon rain forest is directly responsible for the production of 50 per cent of the world's replenishable supply of oxygen.

It is a huge low-lying region consisting of a series of regular tabular formations that descend gradually to the Amazon river. Only 3 per cent of the area consists of alluvial deposits of recent Quaternary origin, the tabular formations being mainly sands and clays of Tertiary age. The basin is flanked by the Guiana plateau to the north and the Central plateau to the south. The basin has an annual rainfall of approximately 2000 mm and annual average temperatures of 22–26° C (Figure 2.5). The highest temperature ever recorded in the Amazon is 36° C.

Essential to the rain forest ecosystem is the Amazon river and its more than 1000 tributaries. Of the twenty largest rivers in the world, ten are in the 5.9 million km^2 Amazon basin. Seventeen of the Amazon's tributaries flow for more than 1600 km.

Figure 2.5
Climate graphs for Manaus and Belem

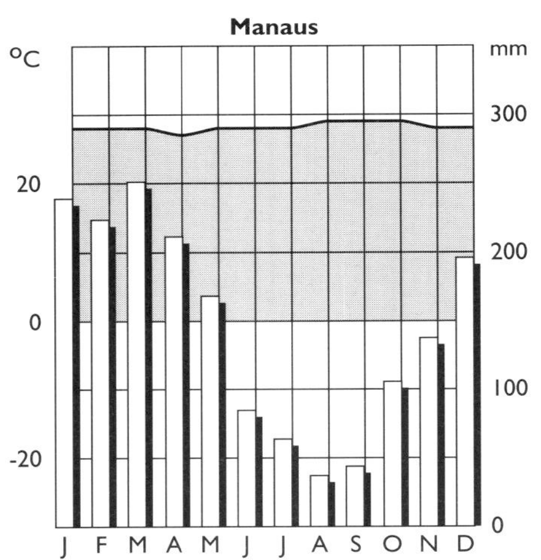

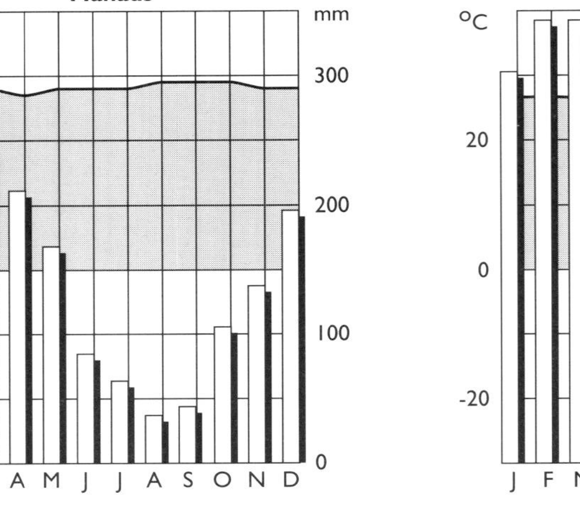

Figure 2.6
Comparative data of major rivers

The Amazon river is second only to the Nile in length (Figure 2.6) and 3615 km of its total length of 6577 km are in Brazilian territory. However the Amazon is the world's largest river in volume and is navigable by ocean-going ships as far upstream as Iquitos in Peru. Dropping less than three centimetres a kilometre after emerging from the Andes, the Amazon drains a sixth of the planet's runoff into the ocean. Its flow is so great that the sea becomes freshwater for 160–325 km out from the shores of Brazil. The force of the river at its mouth is enough to generate waves almost 4 m high. Aquatic plant life on the Amazon is plentiful. The best known are the 'Victoria Regia' water lilies, whose leaves sometimes grow to 2 m in diameter.

Figure 2.7a
Giant water lilies near Manaus

Figure 2.7b The Amazon and its immediate flood plain near Manaus

Figure 2.7c The meeting of the waters of the Solimoes and Rio Negro

Figure 2.7d Recent favela development on the outskirts of Manaus

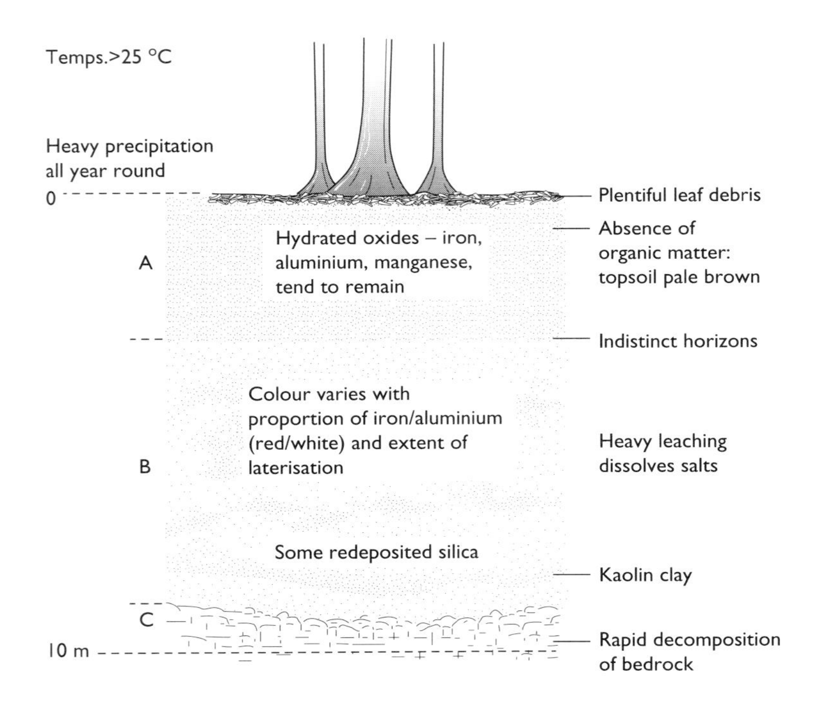

Figure 2.8 Tropical red soils

In general, most of the Amazon basin is under high rain forest on acid soils, known as latosols or tropical red soils (Figure 2.8). These are old and deep soils with a browny-red colour and crumbly texture. Such ferralitic (iron-rich) soils form under conditions of constant high temperatures and moisture surplus, so decomposition and leaching are dominant processes. Decomposition of the leaf litter is rapid because of the high temperatures. Unlike most other rain forest areas, the soils of the Amazon are low in nutrients even when a tree cover exists. The forest receives its nutrients from a network of fine roots close to the soil surface which is mixed with organic material. The root system picks up nutrients directly from the decomposed litter, and nutrients do not get deep into the soil in significant quantity. This organic material is absolutely vital to the ecosystem.

As the parent rock disintegrates rapidly, tropical red soils may become deep if they are formed on level surfaces. With only a moderate gradient they are easily eroded particularly when the vegetation is cleared away.

Most of the rainforest grows on dry land and is known as 'terra firma' rainforest. This contrasts with the lower 'igapo' forest which grows near the rivers on land that is flooded for at least part of the year. Most trees in the 'terra firma' forest grow to between 25 and 30 m, though emergents may be 60 m high (Figure 2.9). The forest is an essential component of the region's water cycle: it has been estimated that about half of the Amazon's rainfall comes from transpiration.

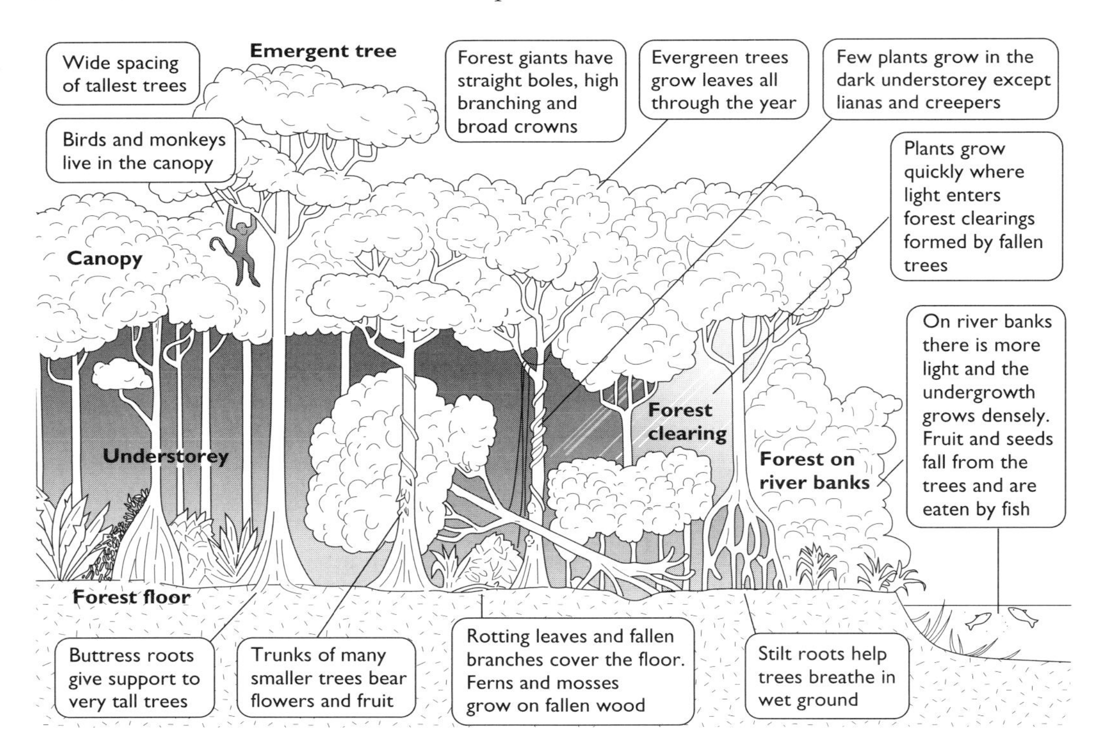

Figure 2.9 Characteristics of the rain forest

EARLY EXPLOITATION

The indigenous inhabitants, such as the Yanomami Indians, lived off the forest for thousands of years without creating any significant environmental stress. Throughout colonial times the fringes of the forest were exploited in a limited way for cattle ranching and sugar plantations but the majority of the region remained untouched apart from an isolated economic cycle based on spices in the north of the Amazon. This was directed by Jesuit missionaries in the seventeenth century who worked among the settled tribal peoples along the rivers. The Jesuits encouraged the cultivation of various crops such as pepper and indigo, for export to Europe. However, in the eighteenth century, the missionaries were ejected from Brazil as from other parts of South America.

The Amazon was the last region of the country to be drawn into the national economy. This integration was based on the exploitation of natural rubber trees in the late nineteenth century (Figure 1.20). Rubber was of limited value until the discovery of vulcanisation in 1839 which considerably improved its properties. Exports of rubber from the Amazon increased from a few thousand kilogrammes a year before vulcanisation to 42 000 tonnes in 1912. Belém and Manaus emerged as the main centres of the rubber trade. The population of Manaus increased from 5000 to 50 000 between 1865 and 1900 while in the latter year Belém's population reached 80 000. At this time nearly a third of Brazil's export trade by value was accounted for by rubber passing through these two ports.

Because of the region's sparse population the rapidly rising labour requirement was met by a considerable migration stream from the Northeast. It has been estimated that between 1870 and 1910, 200 000 people left the Northeast for the Amazon. However the cycle ended rapidly in the early years of the twentieth century with the development of plantation rubber cultivation in south east Asia. In 1905 less than 1 per cent of the world's rubber came from plantations: by 1922 it was 93 per cent. Yet again in Brazil a period of significant economic growth was followed by considerable decline. From then until the 1960s the Amazon was largely ignored in terms of economic development.

RECENT DEVELOPMENT AND ENVIRONMENTAL IMPACT

In recent decades significant areas of rain forest have been cleared. The main reasons have been to:

- provide newly settled smallholders with land for crops;
- create huge cattle ranches;
- build roads;
- exploit mineral deposits;
- use wood for fuel;
- use wood for furniture manufacture and for pulp and paper;
- provide land for urban and industrial uses.

The impact of deforestation, in terms of both Brazil and the planet as a whole, has been well documented, focusing on the major problems of:

1. severe soil exhaustion and erosion;
2. decreasing precipitation due to a reduced vegetation cover;
3. increasing levels of carbon dioxide;
4. reduced releases of oxygen;
5. the adverse impact on the indigenous population;
6. the ultimate alteration of the Amazon's discharge regime and sediment transport pattern;
7. depletion of the region's genetic bank

The Failure of Farming in Rondônia

Population growth has focused on the six main urban areas of Belém, Manaus, Santarém, Cuiabá, Porto Velho and Rio Branco. Outside of these settlements the areas of heaviest growth have been along the Belém-Brazilia Highway and in Rondônia. The latter has become one of the most devastated parts of the rain forest. Migrants arriving along the BR-364 highway were attracted by the apparent fertility of the soil but the fact is that less than 20 per cent of the state is suitable for agriculture. The population grew from around 100 000 in 1970 to over one million in 1990. The catalyst was the paving of the highway in 1984. The failure of farming has had a devastating impact on both landscape and people. Many have turned to mining, often illegally, while the social problems of unemployment have become all too apparent. But in spite of the now well documented problems people continue to move into the region.

The Disaster of Cattle Ranching

In the early 1970s the government decided that significant areas of the Amazon should be cut down to be planted with grass for pasture. A number of foreign multinationals attracted by the idea set up huge ranches. The most suitable breed of cattle for the region, the Nelore from India, required much more grazing land per head in the Amazon than in more fertile environments, up to one hundred times more according to one study which also estimated that over 6 m^2 of forest had to be cut or burnt to produce enough beef for a standard hamburger. In addition, in many areas the quality of pasture declined considerably after a relatively short period. The general demise of the industry in the Amazon eventually came as no surprise.

The Impact of Logging

Amazonia is the world's last great reserve of tropical timber containing an estimated $5 trillion worth of wood. So far much of this has been protected by its inaccessibility. However, since the Belém-Brasília highway was built in 1965 timber companies have flocked to the state of Para. The frontier town of Paragominas with a population of 65 000 now has one of the highest concentrations of sawmills in the world. To quote a recent article from the *New Scientist* on the subject, 'The air is choked with burning sawdust. Throughout the day, truck after truck rumbles by, laden with massive tree trunks.

Figure 2.10 Road construction and forest clearance near Itacoatiara, about 120 km east of Manaus

And even after midnight, the lights are still burning and machines grinding at the countless sawmills that line the road'.

Now the governor of Amazonas is trying to persuade loggers to operate in the relatively untouched western Amazon. Environmentalists are aghast at the prospect. The typical sequence of environmental destruction is as follows:

1. An area of forest is searched for suitable trees to cut.
2. The entangled nature of the forest means that when the selected trees crash down they bring many others with them.
3. More damage is caused when the retrieval team moves in with skidders (bulldozers) to haul out the selected trees.
4. Although a few trees will repopulate the skidder trails, nothing grows where the tyre tracks ran. Trees still fall along the trails, brought down by winds channelled along the newly created corridors.
5. The area is now susceptible to fire for the first time. Normally the closed canopy prevents the moist leaf litter on the forest floor from drying enough to allow any chance sparks to spread. But once the canopy has been broken open fire becomes a very real hazard. Throughout Para fires have wiped out vast areas of logged forest.

The independent research organisation IMAZON has shown that with careful planning and management the impact of logging can be much reduced. In an experimental area:

- vine cutting meant that 30 per cent fewer trees were damaged when felling took place;
- careful route planning reduced by a quarter the area affected by skidders;
- smaller gaps in the canopy and less fuel left behind on the forest floor meant less risk of fire.

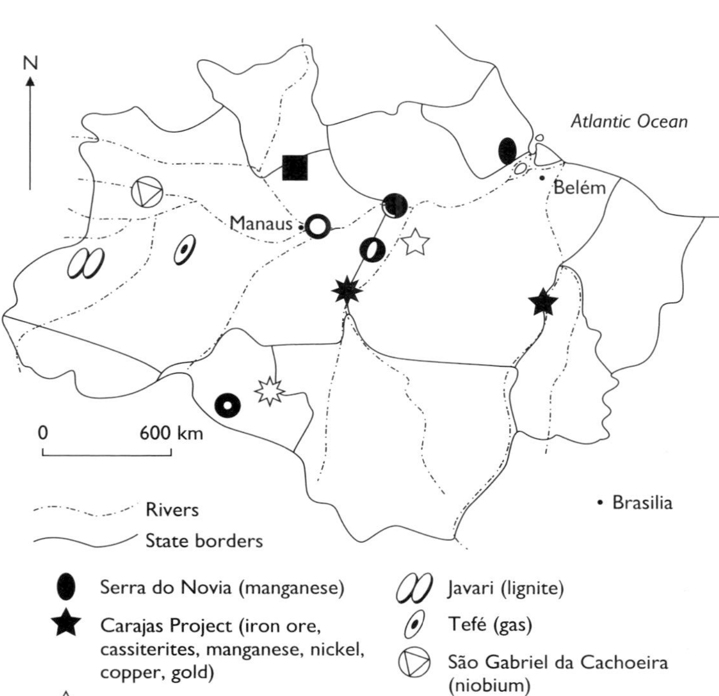

Figure 2.11 Locations of some of the larger mines in the Amazon

The government is making an effort to change the situation. No logging permits are issued by IBAMA, the Brazilian environmental protection agency, without a management plan. However with only 4500 rangers covering a huge area, more often than not management plans are not enforced.

Mining: An Unstoppable Force?

While the economics of arable farming and cattle ranching soon proved largely untenable the same has not been true for mining. Considerable profits have already been made and the mining industry sees much more to come. Figure 2.11 shows the location of the largest mining projects in the Amazon. A consideration of the huge Carajas project can be found on page 50.

GREENPEACE

Greenpeace and other environmental groups have been activily campaigning to preserve the biodiversity of the Amazon (Figure 2.12). Greenpeace, claiming to be the only Environmental NGO acting nationally in Brazil, notes a significant difference in attitude between the Brazilian federal government and some state governments. While the former has been placing environmental issues high on its agenda, a number of state governments, particularly those in the north, have taken the short term financial view, resenting, as they see it, the interference of 'international do-gooders'.

Figure 2.12 Government action to prevent deforestation

ON JULY 26th, Brazil's federal agency for the environment, IBAMA, called the Greenpeace office in Rio. IBAMA's President, Eduardo Martins, took the time to personally inform the Greenpeace office that Brazilian President Cardoso had just signed a decree establishing a two-year moratorium on mahogany logging in Amazonia. 'This is a Greenpeace victory' concluded Martins. After four years of campaigning, the mahogany moratorium is a major breakthrough against the interests of commercial loggers.

In another significant decision, the Brazilian President has increased the protected area in the whole Amazon region. Up to now, 50% of forest regions could be changed to agricultural use. The new act will now limit agricultural use to 20% of the region. In his phone call to Greenpeace Brazil, IBAMA President Martins also asked for Greenpeace support to convince Parliament to approve this provisional act.

But opposition to Greenpeace's goals remains strong. The campaign to preserve Amazonian biodiversity brings Greenpeace up against companies, loggers and politicians from northern states. Deforestation in the Amazon is still on the increase – from 11 130 km^2/year in 1991 to 14 896 km^2/year in 1994.

In future, Greenpeace Brazil will be embarking on a campaign to protect endangered species such as birds, monkeys and other animals which are rapidly losing their habitat as Amazon deforestation continues. The campaign will also focus on international carriers who turn a blind eye to this illegal trade.

Greenpeace Business, Oct/Nov 1996

MEMBERSHIP OF E-8

The Worldwatch Institute, a prominent international environmental agency, recently included Brazil among its group of eight most influential nations in terms of the world's environment. Brazil is grouped with the USA, India, China, Indonesia, Russia, Japan, and Germany in what is known as the 'E-8' – countries whose output disproportionately affects the world's health and whose policies are critical to global climatic change. 'These countries have the Rio agenda as well as the destiny of the planet in their hands', said Christopher Flavin, vice-president of Worldwatch, referring to conventions agreed upon in the Earth Summit held in Rio de Janeiro in June 1992.

Worldwatch say that Brazil has been one of the worst countries in the world in terms of forest devastation in recent years, destroying over 15 000 km^2 per year of its tropical forests between 1991 and 1994. This has led the Brazilian government to declare a moratorium on any new extraction of mahogany and other tropical hardwoods from virgin forestland, and to increase the percentage of forest reserves on new tracts of land from 50 per cent to 80 per cent.

A NEW ENVIRONMENTALLY CONSCIOUS COUNTRY?

Attitudes about the environment have changed markedly in Brazil in recent years, as they have in many other developing nations. There can be little doubt that this increased awareness of environmental matters has been the main stimulus to the introduction of stricter legislation. Because of these changes Brazil now sees itself as an NECC – a New Environmentally Conscious Country.

Much of the credit for raising national awareness about the rain forest must go to INPA – the National Institute of Amazonian Research. Created in 1952, the activities and influence of INPA have expanded considerably in the 1990s. INPA coordinates all Brazilian and much international research in the Amazon region. The organisation's headquarters is in Manaus but it also controls four bases for riverine research and manages research bases in the states of Acre, Roraima, and Rondônia. The Institute also owns two forest reserves and two ecological stations. INPA states that its mission is to 'generate, promote and distribute scientific and technological information about Amazonia for the conservation of its environment and the sustainable development of its natural resources, for the principle benefit of the people who live there'.

SIVAM

Brazil is to put the forests of the Amazon under electronic surveillance in an environmental project unique in size and complexity (Figure 2.13). At a cost of more than $1 billion, satellites, aircraft and ground-radar stations will be linked up to watch over humans, animals and plants. The scheme goes under the name of Sivam, a Portuguese abbreviation for the Amazon monitoring system. However the project has been hit by numerous delays and whether it will survive completely unscathed is currently a matter of debate.

Figure 2.13 Monitoring the Amazon basin

BRAZIL is to put the forests of the Amazon under electronic surveillance in an environmental project unique in size and complexity. At a cost of more than $1 billion, satellites, aircraft and ground-radar stations will be linked up to watch over humans, animals and plants.

The Amazon Surveillance System (Sivam) will cover more than 5 million km^2 an area over half the size of the United States.

Although Brazil claims the destruction of rainforests has slowed by 50% over the past three years, illegal logging continues. Greenpeace, the environmental group, believes that every year smugglers export timber worth $1.2 billion.

To counter this, better use will be made of the Landsat satellite images Brazil currently receives. These will be combined with data from weather satellites to monitor the forests, particularly to watch for fires. Further information will come from eight weather-radar stations and about 50 other weather-monitoring bases.

Sivam will also have an impact on crime fighting, with other airborne and ground-based radars able to monitor drug traffickers in the air, on the rivers or travelling overland. Sensors will keep a check on the quality of river water, where, for instance, an increase in mercury could be a sign of illegal goldmining.

Raytheon, the Amazon defence and electronics contractor, will lead a team of American and Brazilian companies working on the project. One big element will be the integrated telecommunications network needed to handle the vast amount of data from monitoring sources. Because existing communications are limited, much of the data will be sent by satellite or high-frequency radio links.

Brazil will pay for the project under a special deal worked out with the American government. Taking six years to complete, the Brazilians describe it as 'the largest and most complex attempt to monitor the environment anywhere in the world'.

Sivam's headquarters will be in Brasilia, where officials from a number of government agencies collect and process data. According to Raytheon, Brazil will use this to help protect the environment and support conservation while making best use of the land to advance economic development.

There will also be a number of regional co-ordination centres in the Amazon basin: in Belem, a large port near the mouth of the river; in Manaus, 1200 km upriver; and at Porto Velho, close to the border with Bolivia. These centres will be responsible for pinpointing and investigating anything suspicious.

Sunday Times, 31 July 1994

CHALLENGING TRADITIONAL VIEWS

In recent years some scientists, among them geographers, have challenged traditional views concerning global dependence on the rain forests (Figure 2.14). Although the weight of opinion is against this radical school at present, the debate will surely hot up in the years ahead. However, even if the radical view, that we don't 'need' the rain forests, were proved to be right, there are many reasons why the world would still 'want' to keep them.

Figure 2.14 'Little green lies'

'Little green lies'

The North has adopted the rainforests as its special concern, and imbued them with its own values. Whether the countries of the North have done this out of guilt for the clearing of their own forests, or because they wish to extend the physical control of their colonial past through the mind-set of ecology, it is hard to know. Unquestionably, however, the Northern middle classes see the preservation of 'rainforest' as a key item on their global 'stability' agenda. And having agreed this agenda, they have started a process of scientific 'myth-making' to bolster their case. These myths are the 'little green lies' of the current paradigm. The fundamental aim behind them is to make us all believe that we really do *need* rainforests, not just want them or like them.

Most ludicrous of all, however, is the idea that rainforests are '*literally* [our italics] millions of years' old . . . the individualistic response of species to environmental change, and the fossil record clearly indicates that the majority of present-day 'rainforests' are less than 18 000 years old.

There are a series of further scientific 'myths' which are also widely employed to bolster the case for needing rainforests:

1 The rainforests are the 'lungs of the world'. In fact, because of their decomposition processes, most rainforests tend to use up as much as or even more oxygen than they give out.

2 Rainforests are a vital carbon sink, and their cutting and burning will lead to enhanced greenhouse warming . . . when the forest is removed, it is not a case of perfect carbon storage being replaced by no carbon storage. The replacement systems, such as grassland, will also have complex carbon cycles, which need to be taken into account.

3 Rainforests are vital in preventing soil erosion. In some circumstances, this may indeed be true, but many other types of surface vegetation, when properly managed, such as grasslands, can be more effective.

This cavalcade of 'little green lies' has three main purposes, all relating to the essential political ecology of rainforests. The first is the desire of the North to maintain a controlling interest over the resources of the South. The second is the worry that changes in the South, both ecological and economic, might damage the politics of the North. The third is the desperate drive of scientists to obtain continued funding for research into the questions that the current paradigm places on the agenda.

Global Environmental Change
[P.D. Moore, B. Chaloner, P. Stott] 1996.
Blackwell Science

? ? ? ? ? ? ? QUESTIONS ? ? ? ? ? ? ?

1. Write an ecosystem profile of the Amazon rain forest, summarising the climate, vegetation and soils of the region.

2. Assess the importance of the river system to the rain forest environment.

3. How does earlier exploitation of the rain forest compare with its present economic use?

4. To what extent can the forest be viewed as a storehouse of resources?

5. Why has the forest been used as a source of fuelwood?

6. Examine the actual and potential effects of deforestation at the local, national, and international levels.

7. (a) How will the Sivam system operate?

(b) Assess the arguments for and against this expensive monitoring system.

8. Present a detailed explanation of how you feel the rain forest should be managed in the future.

9. Summarise the view now being expounded by some scientists that the planet does not 'need' the rain forests.

The Pantanal – the World's Largest Wetland

Wetlands are among the richest and most productive ecosystems on the planet. They are also one of the most fragile and threatened. Approximately half of the world's wetland systems, particularly those in the northern hemisphere, have been lost forever. Now the largest area of wetland in the world is facing its greatest threat from human activity to date.

Figure 2.15
The Pantanal in the summer wet season

THE PROPOSED SUPER WATERWAY

A £1.3 billion project involving the construction of a 3000 km 'super waterway' to allow ocean-going ships to sail into the heart of South America was announced in 1995. The Hidrovia (Portuguese for 'waterway') project, would start from the Uruguayan port of Nueva Palmira, link the Paraná and Paraguay rivers, cut across the Pantanal and terminate at the Brazilian town of Cáceres in the Mato Grosso (Figure 2.16). The scheme is designed to turn the entire length of the Paraguay-Paraná river system into a modern shipping system. By dredging, damming and diversion a navigable canal would be created, up to 50 m wide and 4 m deep, that would accommodate ships carrying up to 50 000 tonnes. Construction was scheduled to begin in 1997, financed by the four Mercosur countries of Brazil, Argentina, Paraguay and Uruguay, and landlocked Bolivia. The idea, described by the four economic partners as 'the backbone of Mercosur' is not new but was long held up by political and financial disputes as well as technological obstacles.

Ninety years ago one ocean-going ship a day docked in Corumbá, a bustling port on the upper Paraguay. However by the middle of the century commercial traffic on the Paraguay and Paraná rivers had greatly diminished due to (a) the growing dominance of motor transport and (b) the increased size of ocean-going cargo vessels which in general could not negotiate the two rivers due to limited depth and difficult to negotiate meanders in many places.

Figure 2.16 The Hidrovia project

The Economic Rationale

The potential economic benefits of an improved waterway are considerable. It could provide the landlocked countries of Bolivia and Paraguay with an outlet to the sea, create a modern international port in Uruguay, and boost exports from north eastern Argentina. For Brazil, it could trigger an economic boom for farmers and mining companies in the central plains, which are rich in soya beans, wheat, rice, hardwoods, iron, manganese and precious stones. At present, getting these goods to distant markets takes days and sometimes weeks across thousands of kilometres of rough terrain. Ninety per cent of the region's cargo is moved along roads of poor quality. Not surprisingly the price is high: moving grain to the ports of Paranaguá and Santos in south eastern Brazil takes two and a half days and costs \$ 70–110 a tonne. Water transport using the present river system is cheaper at \$ 30–50, but it takes about a month for these vessels of limited size to wend their way from Cáceres to the sea. The Hidrovia scheme would greatly reduce travel time and transportation costs. Proponents of the project believe that it could open up South America's heartland to commerce as the Mississippi river did for the southern USA in the last century.

Figure 2.17 Is the Hidrovia project really a good idea?

Pledges made for largest wetland

NEW environmental guarantees may protect the Brazilian Pantanal, the largest wetland in the world, from the worst impacts of a multi million dollar project to make the rivers Paraguay and Paraná navigable to ocean-going vessels for 3400 kilometres.

Environmental groups, such as the San Francisco-based International Rivers Network (IRN), have warned that the Hidrovia project could cause a catastrophic decline in water levels, leading to the drying out of 40 per cent of the Pantanal's 200 000 km^2.

But the Intergovernmental Committee on Hidrovia issued a statement last week that meets some of these concerns. 'Intervention to make the Pantanal navigable,' it says, 'will be based on signs, buoys, communications' improvements and measures for navigation security, preserving the natural ecosystems.' It also states the need to protect other wetland ecosystems, and guarantees that environmental impact studies 'will be opened to public scrutiny'.

Glenn Switkes, director of the IRN for Latin America, has welcomed the opening of 'a broad and open debate on the Hidrovia'. But the IRN and other groups have yet to be convinced that the Hidrovia project can proceed without serious impacts.

'Hidrovia-related construction even outside the Pantanal could still spell the wetland's demise,' warns Alcides Faria, executive secretary of the 200-member Ríos Vivos coalition. 'The Pantanal's ability to retain its water is regulated by rock formations hundreds of kilometres below its limits, and exploding these to improve navigation, as is currently planned, could cause the desiccation of the Pantanal.'

A further fear is that improved access to wilderness areas will open them up to large-scale export-oriented farming. Paraguay is planning to develop much of the Chaco wetland in the north of the country for agriculture.

Economic, engineering and environmental studies into the Hidrovia are going ahead with finance from the Inter-American Development Bank and the United Nations Development Programme.

The Geographical Magazine, November 1995

Environmental Danger to a Fragile Ecosystem

However there has been considerable environmental concern about the impact of the Hidrovia waterway (Figure 2.17). An article about the project in one scientific journal is entitled 'Hell's Highway' (New Scientist, 3 June 1995). Of greatest alarm is its route through the Pantanal Depression or simply Pantanal (Portuguese for 'swampland') which ranks as one of the most unique regions in the world. Located in the upper Paraguay River basin, this 200 000 km^2 region of wetlands sprawls across Brazil, Bolivia and Paraguay. It is the biggest flood plain in the world and 80 per cent of the Pantanal is within Brazilian territory. Within the region the Paraguay River is joined by many tributaries such as the São Lourenco (670 km) and Cuiabá (650 km) rivers to the north; and the Miranda (490 km), Taquari (480 km), Coxim (280 km), and Aquidauana (565 km) to the south (Figure 2.18).

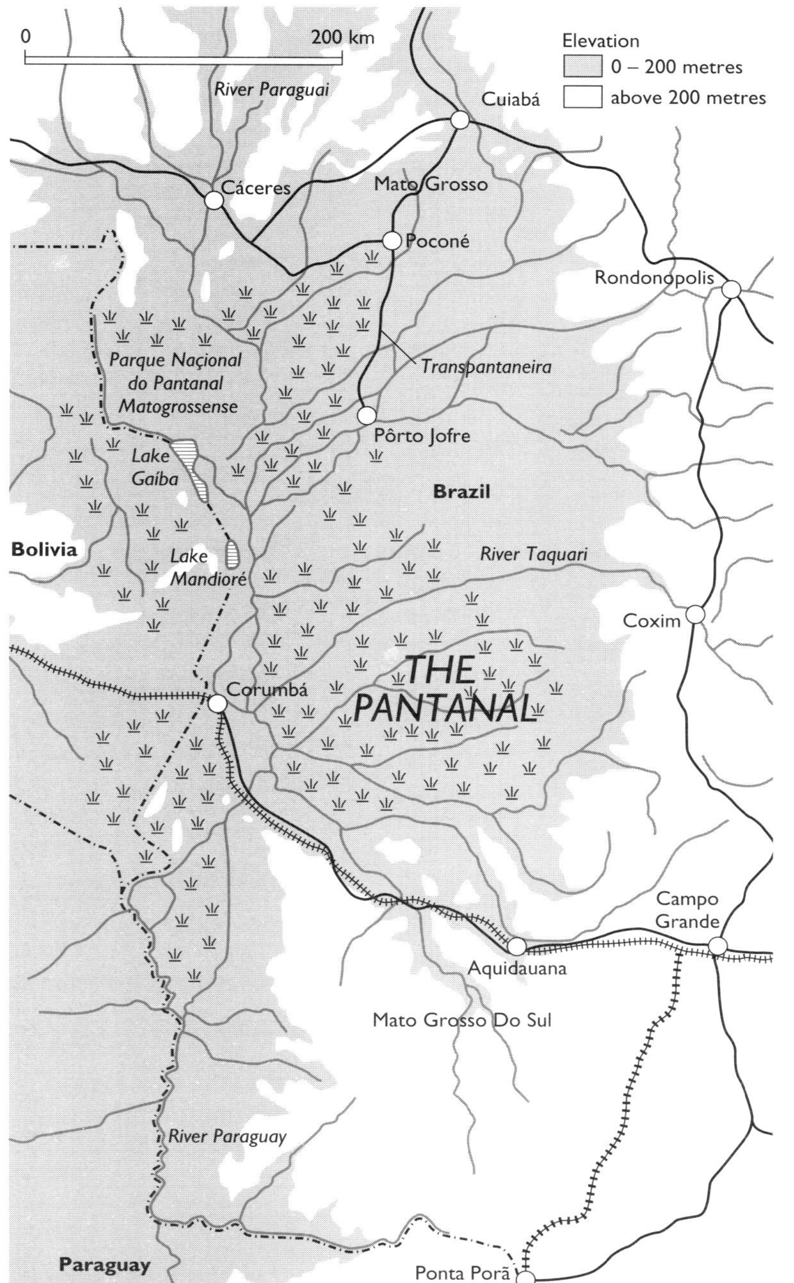

Figure 2.18 Roads and rivers in the Pantanal

The Pantanal is a 'transitional ecosystem' linking the Cerrado in Central Brazil, the Chaco Plains in Bolivia and the Paraguay and Amazon Basins. Particular landscape features are:

- *as lagoas* – temporary or permanent lagoons of various sizes.
- *as salinas* – salty lagoons permanently isolated by ridges of sand.
- *as cordilheiras* – narrow, elongated sand hills covered by savannah vegetation.
- *as vazante* – natural drainage channels which are active during the flooding period.
- *os corixos* – small perennial water courses which link adjacent lagoons.

The gradient of the landscape is only 6–12 cm in an east-west direction and 1–2 cm from north to south, favouring flooding which logically develops from north to south and from east to west.

Figure 2.19 Climate graph for Corumbá

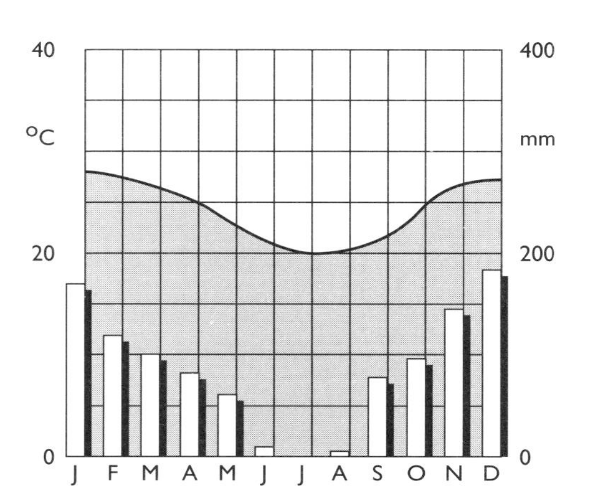

The region is dominated by poor sandy soils with small areas of rich clay and lime soils. The climate is hot during the summer, averaging close to 30°C (Figure 2.19). The annual rainfall of between 1000 and 1400 mm falls mainly during the summer months. Winter temperatures average around 21°C but frost can occur occasionally.

The Pantanal is a sanctuary for an array of flora and fauna, including marsh deer, jaguars, giant ant-eaters, caymans, giant otters, toucans and rare hyacinth macaws, as well as 240 varieties of fish, 50 kinds of reptile, 650 species of birds, 1130 species of butterfly and 90 000 varieties of plants. This homeland to 19 indigenous peoples is without doubt one of the world's biodiversity hotspots.

In summer the Pantanal is a huge lake hundreds of kilometres across as rivers flood their banks. But in winter it largely dries up, leaving swamps, bogs and elevated plains. The high-water mark is usually in February and the waters start to retreat in March. During the dry season, animals gather around the remaining water holes. Only 5 per cent is a national park despite its status as a UNESCO World Heritage Site. This vast alluvial plain, lying between 100 and 200 metres above sea level, is the remains of an ancient inland sea called the Xaraes, which began to dry out 65 million years ago.

Only one road, the Transpantaneira (Figure 2.18), cuts right into the Pantanal. It is a raised dirt road 145 km long from Poconé to Pôrte Jofre. It never reached Campo Grande as originally planned, and with 123 bridges already the cost of building the remaining 202 km is regarded as too high.

Figure 2.20 Traditional farming in the Pantanal

Throughout its history, up to the 1970s, cattle raising was the dominant economic activity in the Pantanal region. 'Pantaneros' (cowboys) ranch 400 000 head of cattle which are constantly on the move from one dry area to the next. In the 1980s the economy of the area began to develop through a diverse set of activities: tourism, professional hunting (mostly illegal), gold and diamond prospecting, road and hydropower-dam building, as well as intensive soy cultivation in the region's southern border. Such activities have created negative environmental impacts on the fragile ecosystem of the region. Adding to the risks derived from economic diversification is the contamination of the rivers flowing into the Pantanal.

On the positive side a specific programme for protecting the region, addressing both local and extra-regional menaces to the ecosystem is being implemented by the federal government in cooperation with State environmental agencies, and local and international NGO's are also very active in the area.

However environmentalists feared that, with the construction of the canal, entire species would be lost and that the Pantanal might even be reduced to a desert. According to the 1993 report by Wetlands for the Americas, the project threatens to disrupt the region's seasonal variations in the levels of water and sediment on which the ecology depends. It could also alter the hydrology of the river basin. The Pantanal acts as a giant sponge, holding water in the rainy season and releasing it gradually throughout the year. This helps to prevent flooding of the lower reaches of the river system where 25 million people live. Leaders of the Indian tribes in the region have expressed their alarm about the possible impact on landscape and their lives.

THE HIDROVIA PROJECT: A RETHINK?

More than 80 campaign groups, scientific institutes and non-governmental organisations, brought together as the Hidrovia Coordinating Committee by local agencies in South America and by the International Rivers Network in San Francisco stated they were 'deeply concerned with the likely social and environmental impacts of the Hidrovia Project'. As a result of mounting concern about the huge potential environmental impact a more conservative scheme based on the clearance of debris, dredging in places, navigation buoys and standardised customs practices, is now envisaged. It will cost in the region of $ 100 million and be far less menacing to the Pantanal, and work will not start before a $ 10 million environmental and economic study is complete. The environmentalists and Indians have won the day for the moment: it remains to be seen how they will react when the final version of the revised plan is published.

Figure 2.21
Life in the Pantanal

Home on an Eco-Range

BRAZILIAN POET MANOEL de barros once counseled against using any kind of yardstick in the Pantanal, for 'the yardstick,' he wrote, 'is the embodiment of limit. And the Pantanal knows no limits.' As the single-engine Cessna crosses the Maracajú escarpment, out over a vast sea of green, it is easy to see what De Barros meant. The Pantanal, a vast South American Everglades in the heart of the continent, stretches to the horizon, crossing into Bolivia 300 km to the west. We are heading to a guest ranch that shares a unique world with fewer than 300 ecotourists a year.

The watery Pantanal is legendary for its exuberant birdlife—some 650 species in all, including roseate spoonbills, red-billed grebe, laughing falcons and giant jabiru storks—as well as some 80 species of mammals and 50 reptiles. In the dry season, which lasts from May to October, the waters recede, exposing white sandy beaches that are prime basking spots for the plentiful caiman, a South American crocodile, and the jaguar.

The morning is spent exploring the marshes and fishing for golden dorado, a South American sport fish. When a catch is hauled into the boat, a torrent of piranha boils in snapping pursuit. During the 20-km ride back upstream, the riverbanks thrum with activity. There is not another human soul in sight.

Every outing at the Black River Ranch, led by a resident biologist-cum-tour guide, offers new surprises—black howler monkeys scampering across the treetops, giant anteaters lumbering through the tall grass. But the real charm of the ranch is the time-worn hospitality and sense of experiencing Brazil's frontier history.

The ranch can sleep 40 guests, but often has only a handful. Despite a surge in ecotourism in the region, the number of visitors at Black River is unlikely to increase. Periodic flooding makes a reliable road link to the outside world virtually impossible. The river is blocked at both ends by impenetrable swamp. There is no way to get in carrying a yardstick.

Time Magazine, 1996

? ? ? ? ? ? ? QUESTIONS ? ? ? ? ? ? ?

1. (a) Describe and explain the proposed route of the Hidrovia waterway.
   (b) Why might this be just the first step in a continental waterways network (Figure 2.16)?
2. Why is the Pantanal such a unique environment in terms of its landscape, flora and fauna?
3. (a) Historically, what has been the economic use of the Pantanal?
   (b) How did human use of the Pantanal change in the 1980s?
4. How would an improved waterway change the economic prospects in its hinterland?
5. Detail the environmental impact that the construction of the waterway according to the original plan might trigger.
6. To what extent does the present level of eco-tourism impact on the region (Figure 2.21)? What are the advantages and disadvantages of developing the tourist industry in the Pantanal?

Brazil as a major producer

A WIDE RESOURCE BASE

Brazil produces a wide range of minerals (Figure 3.1). The country is a major producer of iron ore and niobium and is among the top ten producer countries of aluminium/bauxite, gold, manganese, tin, quartz, kaolin, phosphate rock, chrome, ilmenite, graphite, nickel, rare earths, ferro-alloys, precious stones, asbestos, flourite, magnesite and dimension stones. The foreign trade balance for minerals in 1993 produced a deficit of $ 1 billion, although if petroleum and gas are excluded from the total, there was a surplus of $ 1.2 billion.

Figure 3.1 Brazilian mineral production

	1994
Iron ore ('000 tonnes)	165 651
Bauxite ('000 tonnes)	8280
Coal ('000 tonnes)	4337
Phosphate ('000 tonnes)	3533
Manganese ('000 tonnes)	2321
Dimension stone ('000 tonnes)	2000
Copper	39 674
Tin	22 500
Chromium	147 200
Nickel	16 508
Niobium	14 400
Potassium	230 400
Zinc	187 304
Gold (kg)	72 000
Asbestos	191 900
Kaolin	953 000
Fluorite	76 200
Gypsum	876 800
Magnesite	279 600
Natural gas (million m^3)	7737
Petroleum (thousand m^3)	40 182

(Metric tonnes unless otherwise indicated)
Source: Mining Annual Review 1995

IRON ORE

Iron ore was the first mineral to be exploited this century on a substantial scale. In 1994, Brazil ranked second worldwide among producers (Figure 3.2). In that year 50 mining companies operated 100 surface mines. Exports accounted for almost three-quarters of total production with Japan and Germany being Brazil's main customers. Brazil has the fifth largest reserves of commercially recoverable iron ore, amounting to 230 billion tonnes. Most of the deposits are found in Minas Gerais and Pará, the former holding three-quarters of the total. The Iron Quadrangle, south and east of Belo Horizonte in Minas Gerais has been mined for over 30 years. More recently the high quality deposits of the Serra do Carajás in southern Pará have become the focus of attention.

Figure 3.2
Production of iron ore, 1994

RANK ORDER	COUNTRY	PRODUCTION ('000 TONNES)	% TOTAL
1	China	239 000	24.6
2	Brazil	168 000	17.3
3	CIS	136 000	14.0
4	Australia	128 700	13.3
5	United States	58 400	6.0
6	India	57 500	5.9
7	Canada	37 000	3.8
	World	970 700	100.0

MANGANESE

In 1994 Brazil was the world's fifth largest producer of manganese with 11.1 per cent of the total, most of which is shipped to the steel and ferro-alloy industries. Reserves of 69 billion tonnes also place Brazil fifth in the world. Deposits, in order of importance, are found mainly in Pará, Mato Grosso do Sul and Minas Gerais. These states are consequently the main producers.

BAUXITE

Bauxite production in 1994 was 8.28 million tonnes. Eighty per cent originated from the giant MRN mine in the northern part of Pará state by the Trombetas river. National reserves of 2.8 billion tonnes account for just over 12 per cent of the global total. Outside of Pará, significant reserves are also found in Minas Gerais and Maranhão. Primary aluminium production amounted to 1.18 million tonnes in 1994 with more than half exported.

Figure 3.3
Bauxite arriving by rail at an aluminium smelter

TIN

Production of 15 400 tonnes in 1995 amounted to 7.5 per cent of the world total and made Brazil the fifth largest producer. Reserves of 602 000 tonnes are located mainly in the three northern states of Amazonas, Pará and Rondonia. In 1995, over 80 per cent of all the primary tin produced was exported in ingot form, the main destinations being the USA and Europe.

? ? ? ? ? ? ? QUESTIONS ? ? ? ? ? ? ?

1. How important is the mining sector to Brazil's economy?

2. Analyse Figure 3.1 in terms of

(a) fuels

(b) metallic minerals

(c) non-metallic minerals.

3. What are the disadvantages of a significant mining industry operating in a country?

Mining in the Carajas Region

A MULTI-MINERAL RESOURCE FRONTIER

The Carajás region (Figure 3.4) has been described as a 'Pandora's box' with 18 billion tonnes of iron ore reserves, 150 tonnes of gold, as well as significant deposits of silver, manganese, copper, molybdenum, nickel, cassiterite and bauxite. The iron ore deposits are the world's largest, capable of supplying this important natural resource for five hundred years. The mining complex is in a range of steep hills, the Serra dos Carajás of eastern Amazonia, located between the Xingu and Araguaia-Tocantins River basins in the south of Pará state. It is 550 km southwest of Belém, the capital of Pará.

The immense mineral wealth has attracted development on a huge scale. Apart from the advanced open-cast mining operation, a completely new network of power generation, transport and processing plants has been constructed, with a rail line to the coast connecting with new port facilities and aluminium factories. The latter and the mining complex are supplied with power by the giant Tucuruí hydroelectric scheme.

CVRD AND THE ROLE OF THE WORLD BANK

The site was originally discovered in 1967 by a subsidiary of US Steel, later to be bought out by the Brazilian state-owned Companhia Vale do Rio Doce (CVRD), one of the largest mining conglomerates in the world. CVRD borrowed $3 billion from the World Bank to construct the mine and a 890 km railroad from the Carajás range to the sea. Construction work on the railroad linking Carajás to Ponta da Madeira, near São Luis on the coast of Maranhão, began in June 1978. Mineral production commenced in 1985 and today over 40 million tonnes of the world's purest iron ore are exported from the site each year, as well as manganese, gold and soon, copper.

Figure 3.4
Location of Carajás

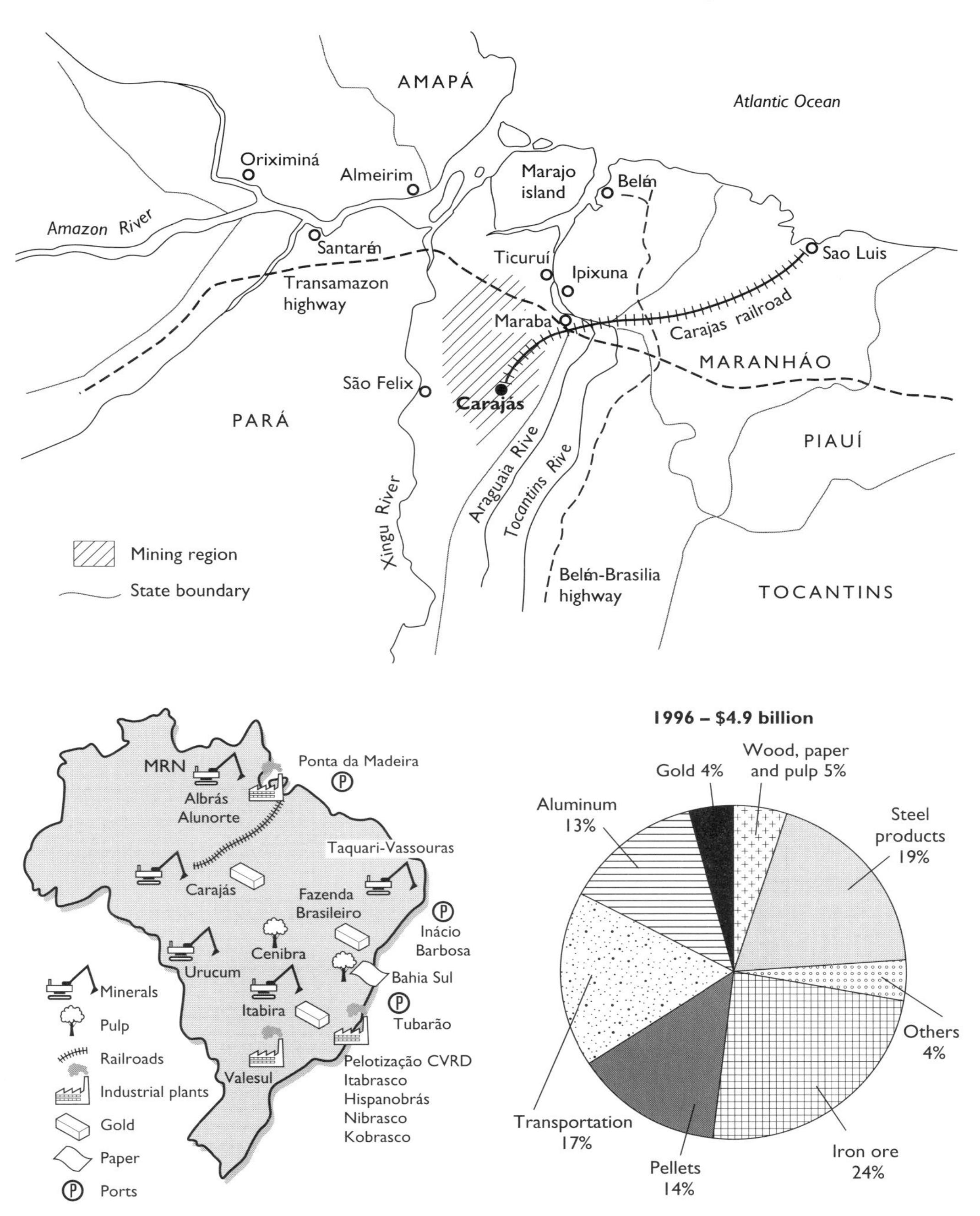

Figure 3.5a CVRDs operations in Brazil **Figure 3.5b** CVRDs gross revenues

Figure 3.5 shows CVRD's current operations in Brazil which may well expand now that the company has been privatised. The Serra dos Carajás region covers 1.2 million hectares, of which CVRD holds the mining rights to 412 000 hectares.

Raw material extraction has regained importance in development strategies as the International Monetary Fund and the World Bank encourage many developing countries to postpone hopes of large scale industrialisation and rely instead on increased exports of raw materials to meet debt repayments and trigger economic growth, a strategy widely rejected just twenty years ago. While Brazil aims to follow both strategies and to benefit from the important links between the two, the Carajás project was clearly designed to raise hard currency in the face of the country's looming debt crisis.

THE GREATER CARAJAS PROGRAM

The 1974 POLAMAZONIA regional development plan sought to combine public and private capital in 15 foci of mining, lumbering and large agro-export plantations. Comparative advantage was the principal consideration in selecting which areas of the Amazon to develop first. Not surprisingly Carajás headed the list and it became the 'growth pole' in plans for a much larger 'Greater Carajás Program', a massive export scheme covering 11 per cent of Brazil, an area larger than Britain and France combined. The project was marketed as a chance to 'integrate' development around the 'growth pole' of mining with significant industrialisation as one of the clear benefits. However, government incentives have mainly favoured the mine and other export-oriented activities along the railroad. These activities include lumbering, ranching, oil palm plantations and the production of pig iron and ferro-manganese using charcoal from the Amazon forests.

The Greater Carajás Program included plans for a series of 20 to 30 pig-iron plants along the railway line but to date only a few have been built. Two are located in Maraba and two in Acailandia but concerns over economic viability and the environmental impact of producing huge quantities of charcoal from Amazon forests look likely to greatly restrict further development.

IRON ORE MINING

The main Carajás iron ore mine is a massive excavation hole 4.1 km long, 300 m wide and 400 m deep. It has been dug out in 15-metre-high step-like benches. Thirty-eight gigantic, bright yellow, off-highway trucks, including 240-tonne Haulpaks, the largest trucks ever built in the world, overcome 10 per cent inclines to deliver 370 000 tonnes of ore and waste each day to the nearby crushing and processing plant. The mining method is state-of-the-art open pit benching with all operations computer guided from hi-tech control rooms. The ore has an average 66 per cent iron content, which is extremely high. Figure 3.7 shows the destinations of Carajas iron ore.

Figure 3.6 The Carajás iron ore mine

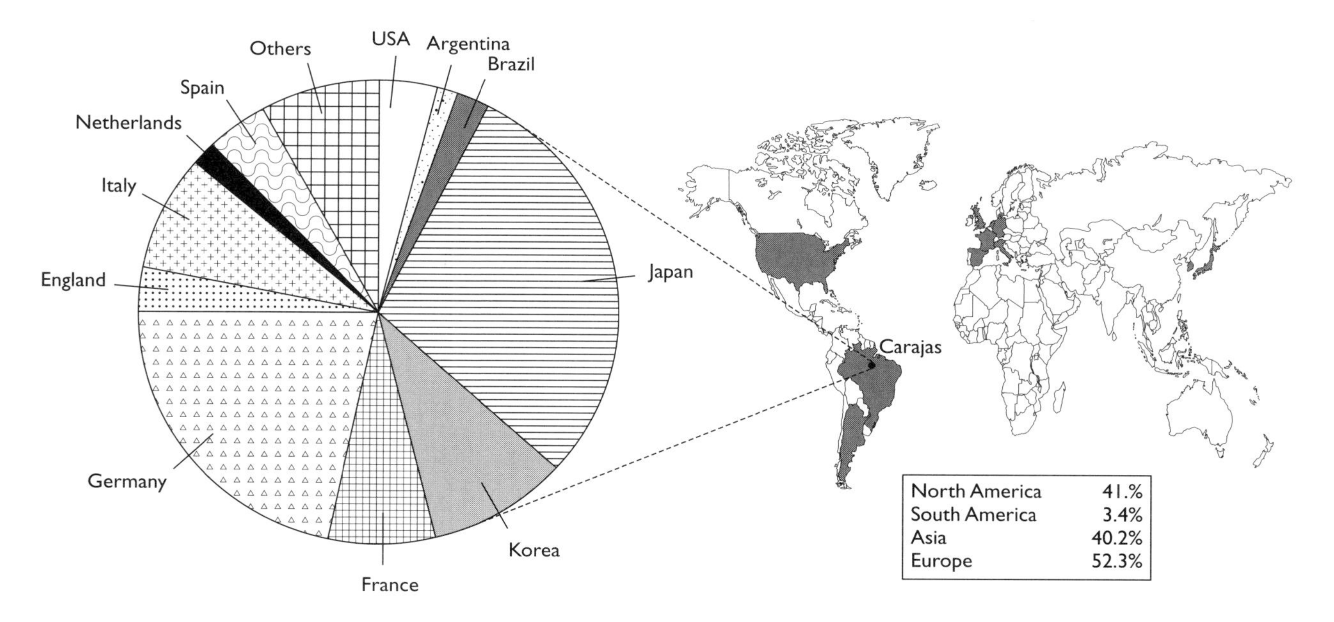

Figure 3.7 The destination of Carajás iron ore

OTHER MINERALS

Just over 100 km from the Carajás mine 5.2 tonnes of gold are produced each year from the Itagape Bahia mine. This open pit operation is one of the most profitable in the world with a remarkably high recovery rate of 93 per cent. The recovery rate refers to the total amount of gold which is actually recovered, with the rest lost in the process of mining, crushing and refining. The site is estimated to hold 14 million tonnes of gold ore, corresponding to reserves of 66.5 tonnes of gold. The ore is hauled from the mine pit by giant trucks to the crushing and refining base nearby. From there the refined gold bars are transported by heavily guarded helicopters to Carajás airport and from there taken by plane to São Paulo to be sold on the market, mainly to banks.

Early in 1996 the government announced the largest gold discovery in the history of Brazil. On CVRD prospecting rights territory, the find is located in Serra Leste, a large escarpment traversing the eastern flank of the Amazon basin. It has been estimated that 150 tonnes of gold lie in deeply buried veins 400 metres or more below the surface. Requiring an investment of $250 million the mine will be the largest gold mine in Latin America when it comes on stream in 2000. Serra Leste also has significant deposits of palladium which is used in industrial alloys and is more valuable than gold.

The area is being contested by individual prospectors, or '*garimpeiros*', some of whom have clashed violently with CVRD personnel. For centuries *garimpeiros* have panned gold in frontier areas and in 1988 they produced a record 90 tonnes, helping Brazil to become a major producer. However, increasingly hand shovels have given way to huge earth-moving vehicles and in 1995 *garimpeiros* found only 20 tonnes of gold while industrial mines totalled 43 tonnes. The Serra do Leste gold will require deep shaft drilling which in reality will leave the *garimpeiros* out in the cold.

At the giant Salobo Copper Project which is due to come on stream in 1998, some eight tonnes of gold is expected to be scooped up as well as a bi-product. Copper reserves in the region have been estimated at 1.2 billion tonnes with a 0.84 per cent metal content. Manganese reserves have been estimated at 65 million tonnes while nickel ore reserves are assessed at 45 million tonnes.

THE GROWTH OF SETTLEMENT

The development of the mine and its associated activities has significantly changed the population structure of the region. The European-origin population of the Carajás region, which was negligible before 1980, grew explosively as work in the construction of the remote project employed up to 25 000 men at its peak in 1983–5. A strange urban dichotomy is apparent, the Carajás residential township which houses 7000 people (key employees and their families) is a twenty minute drive from the mine. It is a well planned model settlement. Facilities include a fully equipped sports centre, a 2000 seat theatre, two fully air-conditioned hotels and the Carajás Zoo-Botanical Park.

Startled by the explosive influx and invasions of miners and colonists into its land, CVRD and the federal government established in 1984 and 1985 a ring of agricultural smallholder settlements around the project to stabilize land tenure. These settlements, called Carajás II and III, were funded and run by the federal government land-colonization agencies GETAT and INCRA in the early 1980s. Each farmer was provided with 100 hectares, materials for a house, and several months' minimum wage. These farmers seemed to be in an excellent position to furnish Carajás with its basic food needs. With heavy seasonal rainfall and almost no maintenance, however, roads quickly disintegrated. Malaria, slow titling, poor agricultural extension, abysmal schools, and a shortage of available credit combined with the difficulty in transport in forcing most colonists to abandon their lots within the first few years. Many moved to other locations or to town, most selling off their partially cleared properties to ranchers.

World Development, Vol 23, No 3

Figure 3.8 The failure of the settlements at Carajás I and II

Outside the controlled-access company town and ecological reserve an uncontrolled jungle boom town named Parauapebas developed. By 1991 the population of Parauapebas had risen to over 30 000, more than four times the size of the company town. The population explosion has strongly skewed the sex ratio toward men and resulted in an influx of prostitution and gambling. However, in 1988 Parauapebas gained sovereignty as a new municipality and control of substantial revenues in taxes from the Carajás mines.

Natural resource boom towns tend to go through four phases: discovery, rapid growth, maturation when resources are concentrated in fewer hands, and a bust phase when reserves of the mineral or its price begin to drop and people move out. The maturation phase has already begun in Parauapebas.

Agricultural smallholder settlements were also established in the region to stabilise land tenure and to increase local food supply (Figure 3.8).

THE CARAJAS RAILROAD

The 892 km Carajás Railroad, originally constructed to transport millions of tonnes of iron ore from the Carajás mine to the Ponta da Madeira sea terminal near São Luis, has expanded its operations. In 1995 it carried an extra four million tonnes of other commodities apart from 44.7 million tonnes of iron ore. The other products consisted mainly of steel, fuel, timber, grain, vehicles and fertilisers. The movement of soybeans has increased rapidly in recent years. In fact without the railway soy production would not be able to reach its markets. It looks likely that the railway will continue to extend its sphere of influence in years to come.

Passenger transport has also risen sharply from 495 000 passengers carried in 1991 to 840 000 in 1995. With several stops between Carajás and São Luis the price of a railway ticket is less than half that of the intercity buses along the same route. Along the line several housing developments and service centres have been built by CVRD to cater for the railroad's 1540 directly employed staff.

Figure 3.9
The Carajás railway through Pará state

ECONOMIC AND ENVIRONMENTAL CRITICISM

The Carajás project has been criticised by some people on economic grounds and by many more because of its environmental impact. The economic criticisms are:

- the huge capital investment has generated a relatively modest number of permanent jobs and stimulated only limited local development. The Brazilian economist Oliveiro Filho wrote in 1988 'One fundamental characteristic of the grand projects of the Amazon is that they capitalise little, not creating significant multipliers in the local economy';
- too much of the area's output comes directly from the sale of a few minerals controlled almost exclusively by one company.

However, it is environmental concern over huge open pit mining in this tropical rain forest region that has caused the most debate. The main issues are:

- a large section of rainforest, the home of several thousand Indians has been transformed into an enormous industrial park;
- against the wishes of the local community the railway line to the coast was built through the Gavioes Indian reserve;
- hundreds of Indians have died as a result of 'imported' diseases;
- in the Xikrin Indian reserve independent miners have invaded and polluted local rivers with mercury, used to separate out gold after panning;
- deforestation has led to the inevitable consequences discussed on page 37;
- air pollution has increased significantly as the scale of economic activity has expanded.

Stung by vociferous opposition in the early years of development CVRD created its Advisory Group for the Environment in 1980. The group included a number of the country's top scientists. In its promotional literature the company state that most of the advisory group's recommendations with regard to Carajás have been totally adopted. CVRD also highlights the environmental conditions imposed on it by the World Bank, one of the most important of which is that all exhausted open pit areas are subject to full reforestation.

? ? ? ? ? ? ? QUESTIONS ? ? ? ? ? ? ?

1. Describe the location and environment of the Carajás region.
2. To what extent is Carajás a multi-mineral region?
3. Suggest why the involvement of the World Bank was crucial to the implementation of the project.
4. Comment on the diversity of CVRD's operations in Brazil and its sources of revenue.
5. What impact has mineral development had on settlement in the region?
6. Assess the importance of the Carajás railway and the port development at Ponta da Madeira to the exploitation of the region's mineral wealth.
7. Examine the environmental impact of the various components of the project. Do the costs of this development outweigh the benefits?

4

ENERGY

An Overview

The availability of cheap, imported energy provided the basis for the economic transformation of Brazil, which began in the 1950s and accelerated during the years of the 'economic miracle'. However, the substantial rise in the price of oil on the world market during the 1973–4 'energy crisis' initiated a drive for energy self-sufficiency, with hydro-electric schemes, the production of sugar-based fuel alcohol, the development of nuclear resources and the increasing exploitation of domestic oil and gas resources. All of these initiatives reduced the import bill for energy but because foreign finance underpinned such development it brought the sector overseas debts of around $25 billion.

Figure 4.1 shows the domestic supply of primary energy in 1994 with renewable energy sources accounting for 61 per cent of the total. Imports remain significant in the national energy balance with almost half the oil required coming from abroad. Much of the coal used in Brazil is also imported. Electricity imports come mainly from Paraguay's share of production from the giant Itaipú HEP plant.

Brazil produces less than 20 per cent of the coal it consumes with a total of less than 5 million tonnes a year. Ninety per cent is steam coal and 10 per cent of metallurgic class. The country's coal reserves are estimated at 1 billion tonnes, located in three fields in the South. Though the coal is of high sulphur and ash content, the government has encouraged its use as an industrial fuel, especially in the cement industry.

Figure 4.1 Domestic supply of primary energy, MTOE*

	1990	1994
Total	187.3	210.9
Renewable	116.4	128.6
Electricity	67.6	79.6
Wood & charcoal	28.2	24.1
Cane bagasse	18.5	21.8
Other	2.1	3.1
Non-renewable	70.9	82.3
Oil	56.6	66.1
Natural gas	4.2	5.0
Coal	9.5	11.2
Uranium	0.6	0.0

(*) In MTOE, a unit of measure equaling 1 000 000 tonnes of oil equivalent
Source: 1996 Guide to the Brazilian Economy

Uranium deposits in Brazil are estimated at about 200 000 tonnes. On this basis, in the 1970s, the provision of nuclear power was thought necessary to meet the growing demand for electricity forecast for the Southeast. An ambitious agreement was signed with the former West Germany for the provision of a total of eight reactors, completion of the uranium ore plant in Pocos de Caldas, and the Pastille fuel plant in Resende. The 626 Mw Angra 1 reactor, located at Angra dos Reis in Rio de Janeiro state, came on stream in 1985 but has operated erratically ever since. Work on the 1300 Mw Angra 3 plant which was due to come on stream in 1992, was halted in 1985 along with two other plants, Iguape 1 and 2.

Investment in Brazil's nuclear energy programme recommenced in August 1995 when the government opened the bids for contracts worth $150 million on the 1300 Mw Angra 2 reactor. Work first began on the 70 per cent complete Siemens-built reactor in 1976, and stopped in the mid-1980s during the debt crisis. The reactor has so far cost a record $4.6 billion, including $1.7 billion in interest payments on loans.

Figure 4.2 Electricity generation in Brazil, gigawatt hours, 1992

Hydroelectric	217 782
Thermal	14 454
Nuclear	1446
Total	233 682

Figure 4.3 Comparative tariffs, July 1995

COUNTRY	TARIFF (US$/MWH)
Japan	202
Portugal	161
Italy	156
Germany	144
Spain	130
UK	127
Chile	113
Argentina	98
Uruguay	87
USA	85
Turkey	80
Australia	76
Brazil	66

Hydroelectric sources dominate total electricity generation (Figure 4.2), accounting for over 93 per cent of the total in 1992. The large scale exploitation of such a valuable indigenous resource has allowed Brazil to price its electricity at a relatively low level (Figure 4.3).

Broad changes in Brazil's economic policy have begun to restructure an energy sector heavily controlled by national, regional, and local governments. In mid-1995, the federal government adopted legislation permitting expanded private ownership and the investment of foreign capital in the country's publicly owned utilities.

Substantial gas reserves and projected huge demand add up to a major emerging market for the gas industry in South America. In the so-called 'Southern Cone' the major producing nations, which are eager to export this valuable commodity are Argentina, Bolivia, and Peru.With its large population Brazil is clearly perceived as the main potential market. However a more highly developed pipeline network will be required to facilitate such transfers (Figures 4.4 and 4.5).

Figure 4.4 The plans for a Bolivia-Brazil gas pipeline, with help from British Gas

Go-ahead for giant gas plan

British Gas is taking part in Latin America's biggest energy project, a $1.8 billion pipeline that will bring natural gas from Bolivia to the industrial heartland of Brazil.

It is an enormous undertaking. The pipeline will stretch for more than 3000 km from Santa Cruz, on to Campinas, and then Port Alegre. Once it is completed, Brazil expects to buy eight million cubic feet of gas a day from Bolivia, rising eventually to 30 million cu ft.

Construction of the pipeline should be underway next year and the first gas is expected to come on stream in late 1998. "It's a tight timetable," says a British Gas spokesman, "but it's something to aim for."

Brazil's energy demands are expected to double over the next decade. Fortunately, gas is one of the most efficient of fuels as well as being one of the cleanest, which is very important in a country that attracts the close attention of environmentalists.

Steel pipes 32 inches in diameter will carry the gas through some of the most difficult terrain in the world. The pipeline will serve more than 70 southern Brazilian cities; the Sao Paulo metropolitan region, with 16 million inhabitants, will take about half the total gas supply.

It is thought around 500 companies will switch to natural gas for their power requirements, with engineering, chemical and petrochemical plants among the most likely users.

British Gas will be particularly involved in the smaller pipework closer to the urban areas. Its spokesman says that "it is a massive project, and we are very proud to be involved in it. There is fierce international competition these days over projects like this".

The pipeline project is British Gas's first entry into the Brazilian market. The company now has offices in Rio de Janeiro and São Paulo with about 12 staff. However, it has been operating in South America for seven years, and is a major gas distributor in the Argentinian capital Buenos Aires.

"We want to be involved in the development of the market downstream," the spokesman for the company adds. "The pipeline is a stand-alone project and, although we shall be involved in that, our main expertise is in the smaller pipelines and distribution."

"In Brazil there is very little use of natural gas and we will want to make gas the preferred fuel."

The Times, 9 December 1996

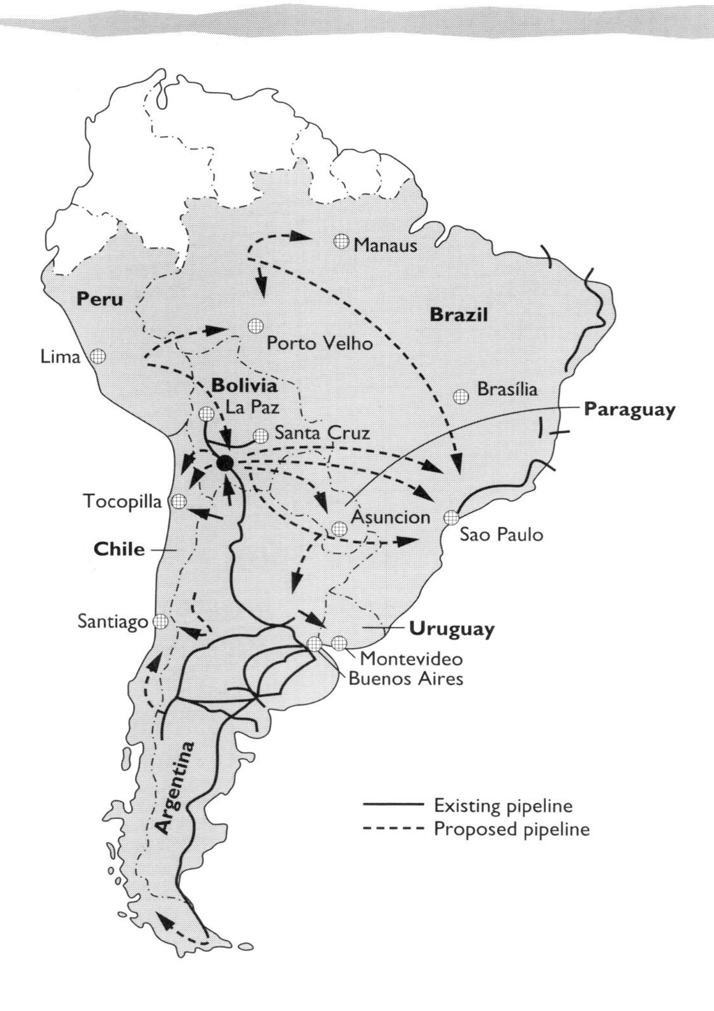

Figure 4.5
Southern Cone gas pipeline systems

? ? ? ? ? ? ? QUESTIONS ? ? ? ? ? ? ?

1. Study Figure 4.1.

(a) What is meant by the term 'primary supply'?

(b) Discuss the significance of each source to total primary supply.

2. Look at Figure 4.2

(a) What proportion of Brazil's electricity is accounted for by HEP?

(b) Explain the limited contributions of (i) thermal power and (ii) nuclear power to total electricity generation in Brazil.

3. Figure 4.3 shows the price of electricity in a selection of developed and developing countries.

(a) How does the price of electricity in Brazil compare to that in the USA, the UK, and Japan?

(b) Suggest reasons for such a variation in price.

4. (a) Assess the significance of the Bolivia-Brazil gas pipeline.

(b) Describe and attempt to explain the existing and proposed gas pipeline networks in the 'southern cone ' region of South America (Figure 4.5).

THIRD LARGEST CONSUMER OF HEP

Brazil ranks behind only the United States and Canada in total annual consumption of hydro-electricity. HEP accounted for more than 97 per cent of the 250 million megawatt-hours of electricity Brazilians consumed in 1994. The country's HEP generating capacity has increased substantially in recent decades from 9100 MW in 1970 to 53 000 MW in 1994. According to data from Eletrobras (1994), Brazil's total hydropower generating potential is estimated to be 127 867.6 MW/year, of which 24.42 per cent are in operation and/or under construction, 35.80 per cent inventoried and 39.78 per cent estimated (Figure 4.6).

In terms of operating hydroelectric capacity, the Paraná basin is the most highly developed by far. But the greatest potential clearly lies in the Amazon basin.

Figure 4.6 Area and hydro-electric potential by hydrographic basins

HYDROGRAPHIC BASINS	AREA KM²	HYDROELECTRIC POTENTIAL (CAPABILITY - MW/YEAR) TOTAL	IN OPERATION AND/OR UNDER CONSTRUCTION	INVENTORY OF BASIC POTENTIAL VIABILITY	ESTIMATED
Total	8 547 374.7	127 867.6	31 227.2	45 780.0	50 860.4
Amazon	3 904 392.8	53 969.7	176.8	16 799.4	36 993.5
Tocantins	813 674.1	14 346.8	3527.2	9284.2	1535.4
South Atlantic,					
Northern and Northeast sections	990 229.1	1579.8	150.5	100.3	1329.0
Eastern section	572 295.8	7392.1	1075.6	5031.0	1285.5
Southeastern section	223 810.2	4415.4	1197.2	1287.2	1931.0
São Francisco	645 067.2	10 379.2	6064.9	3058.8	1255.5
Parana	1 220 411.7	29 361.2	18 894.1	5182.4	5284.7
Uruguay	177 493.8	6423.4	140.9	5036.7	1245.8

VAST POTENTIAL FOR FURTHER DEVELOPMENT

In 1991, the state-owned power sector holding company, Eletrobras, launched a major revision of Brazil's plan for electric power development. The resulting National Electric Energy Plan 1993/2015, commonly called Plan 2015, announced that new projects would be analysed against a broader set of criteria than in the past. This included measurement of all the costs and benefits of various generating sources and full involvement of local social, environmental, and economic development interests in the expansion process. Under the plan, Brazil's 35 principal power utilities will coordinate power system operations more closely to maximise production from available resources and minimise effects on natural streamflow conditions.

Brazilian energy planners believe the outlook for future growth in HEP generating reserves is as bright as its past. Eletrobras estimates that annual electricity consumption will more than double from current levels to around 560 million MWh by 2015. HEP will almost certainly provide the greater part of the generation needed to meet that additional demand. Eletrobras has identified about 200 000 MW of additional HEP potential that could be developed with current technology. About half of the available hydro potential lies in northern Brazil in an area covered by the Amazon rain forest and located substantial distances from major load centres.

Figure 4.7 Comparison of Brazil's primary electric power resources and expected costs of development

	POTENTIAL (MW)	COST (US$/MWH)
Hydroelectric	81.5	<$40
	96.3	$40–$70
	69.2	>$70
Coal	18.0	$60–$70
Nuclear	25.0	$60–$70

Current generation plans include 40 projects with total capacity of more than 30 000 MW that are scheduled to be completed by 2015. Although virtually all of the projects have passed at least preliminary engineering evaluation, only some of this capacity is feasible when issues such as construction costs, power system economics, and environmental factors are considered. Figure 4.7 shows Brazil's primary electric power resources and the expected costs of development. The most recent update of Plan 2015, completed in April 1994, proposes to bring 29 600 MW of hydro capacity on-line between 1995 and 2005.

Brazil has the third greatest reserve of untapped hydropower potential in the world. Not only is the nation's total hydropower capacity impressive, but year-to-year variations in available power are very limited. This stability has made it possible for Brazilian utilities to provide one of the highest quality electrical systems in the developing world.

GLOBAL WARMING

However a recent study carried out by the Brazilian National Institute for Research in Amazonia and published in the journal 'Environmental Conservation' claims that HEP dams with reservoirs in tropical forests can contribute far more to global warming than fossil fuel power plants (Figure 4.8). The research team leader has calculated that in 1990 emissions of CO_2 and methane from water and rotting vegetation in the Belbina reservior had 26 times more impact on global warming than emissions from coal-fired plants generating the same amount of electricity. While the gases released from Balbina will slowly fall as the vegetation decays, they will always be higher than those from equivalent fossil fuel generation. In contrast, emissions from the Tucuruí dam's reservoir had only 60 per cent as much impact on global warming in 1990 as an equivalent coal-fired plant because it floods far less land per unit of electricity generated. The study notes that the total area of reservoirs planned in the region is about 20 times the area existing in 1990, an expansion that will contribute significantly to global greenhouse gas emissions.

Dr Fearnside's calculations have been seized upon by the International Rivers Network (IRN), the US-based lobby group which led criticism of the World Bank's now abandoned plans to build run-of-river dams in Nepal. IRN's Patrick McCully notes that the pro-dam lobby has been shifting its arguments over the years. At one time large dams were advocated because they provided cheap power, despite the social and surface environmental damage caused. More recently large dams have been promoted as 'carbon free', a claim that is now under intense pressure.

Figure 4.8 Is HEP as clean as it seems?

From China to Norway, new hydroelectric schemes are supposed to help cut emissions of greenhouse gases. But will they?

SHIMMERING waters in remote mountain reservoirs providing hydroelectricity seem a world away from the grimy coal-burning power stations of industrial landscapes. Hydroelectric plants burn nothing and apparently cause no air pollution—no acid rain, no grimy particles. They are, surely, the ultimate clean sources of energy. Most countries categorise hydropower as a ''zero-emission technology''. And governments around the world are committed to cutting their future output of greenhouse gases by building hydroelectric dams instead of fossil-fuel power stations.

But their efforts may be misguided. Studies on reservoirs from the Canadian Arctic and the jungles of the Brazilian Amazon are starting to show something odd. Many of these large bodies of water themselves produce greenhouse gases.

One of the worst offenders is the Balbina reservoir on the River Uatumã, a tributary of the Amazon deep in the Brazilian rainforest. The dam supplies most of the electricity for Manaus, the capital of the Amazon region. It is a modest 50 metres high but floods 310 000 hectares of river valley, an area the size of Lancashire or Luxembourg. From the air, the great expanse of water looks more like a temporary flood than a permanent reservoir. A third of the water is less than 4 metres deep and the skeletons of dead and dying trees break the surface almost everywhere.

In common with many other dam builders, the engineers who built the Balbina dam in the mid-1980s made no effort to remove the trees before flooding the site. With no accessible market for timber, felling was not economically viable, so the reservoir was allowed to inundate more than 100 million tonnes of vegetation.

Reservoir bogs

Philip Fearnside, an ecologist at Brazil's National Research Institute for Amazonia, has studied the gases emitted from the reservoir. Decaying plants breaking the surface water produce CO_2, as does rotting vegetation in the oxygenated waters within a metre or so of the surface. But vegetation in the anaerobic waters produces methane. In shallow areas of the Balbina reservoir, says Fearnside, ''methane bubbling can be seen everywhere''.

Fearnside calculates that in the nine years since the Balbina reservoir formed behind its dam, there has been a dramatic release of CO_2. In 1988, the first year after flooding, the reservoir emitted more than 10 million tonnes. Today, the annual figure has fallen to about a quarter of that. Methane is produced much more slowly. In the first year, some 150 000 tonnes emerged from the reservoir. But that figure will be maintained more or less indefinitely. In the Amazon, says Fearnside, it takes 500 years for a tree trunk to decay in anoxic water.

Assessing the damage resulting from the release of CO_2 and methane is complicated. CO_2 persists much longer in the atmosphere—an average molecule will stick around for a century, compared to a decade or so for methane. But while it is there, methane is a much more potent greenhouse gas. So what time frame should be used for any damage assessment? Fearnside adopts a formula from the UN's Intergovernmental Panel on Climate Change that, for the first century after its release, a molecule of methane will have 11 times as powerful a greenhouse effect as a molecule of CO_2. A longer time frame produces a lower figure. A shorter period produces a higher figure. Over the first 20 years, for instance, methane is 60 times as potent.

Dam pollution

At Balbina, Fearnside has aggregated the greenhouse effect of methane and CO_2 to give a ''CO_2 equivalent'', that he can compare directly with emissions from fossil-fuel power stations. He calculates that Balbina emitted the equivalent of more than 12 million tonnes of CO_2 in its first year. That figure fell to some 7 million tonnes in 1990 and to around 2 million tonnes last year. It will slip below 1 million tonnes in about 10 years' time, and drop to 0.5 million tonnes in perhaps 50 years.

If the Balbina dam had not been built, the authorities in Manaus would probably have constructed a conventional power station burning diesel and fuel oil. Such a plant would have produced annual emissions—almost all of it carbon dioxide—of some 0.4 million tonnes, says Fearnside. So far, Balbina has had something like 16 times as potent a greenhouse effect as an equivalent fossil-fuel power station. And, says Fearnside, despite gradually reduced emissions, it will continue to be more polluting ''for 50 years, and probably indefinitely''.

New Scientist, 4 May 1996

? ? ? ? ? ? ? QUESTIONS ? ? ? ? ? ? ?

1. Draw a percentage bar graph to show the division of Brazil's hydroelectric potential in term's of (i) in operation/under construction (ii) inventoried (iii) estimated.
2. With the aid of an atlas assess Brazil's hydroelectric potential by hydrographic basin.
3. Comment on the country's current generating plans.
4. Analyse the data presented in Figure 4.7.
5. (a) Discuss Dr Fearnside's assessment of the relationship between HEP dams and global warming .

   (b) What impact might this research have on the future development of HEP in different parts of the world?

Offshore Oil and Gas in the Campos and Santos Basins

OIL: INCREASING PRODUCTION BUT STILL IMPORTING

In 1995 it was estimated that Brazil's exploitable reserves of oil were approximately 8.9 billion barrels, of which 6.2 billion were considered to be proven resources and 2.7 billion to be probable. The bulk of these resources were discovered in and since the second half of the 1970s. The reserves are concentrated in 12 offshore sedimentary basins (over 1 million km^2) and three major onshore basins (over 3 million km^2). The offshore fields account for over four-fifths of known reserves. With 1.1 per cent of global production in 1995 Brazil was 22nd in the world oil ranking, but second only to Venezuela in South America.

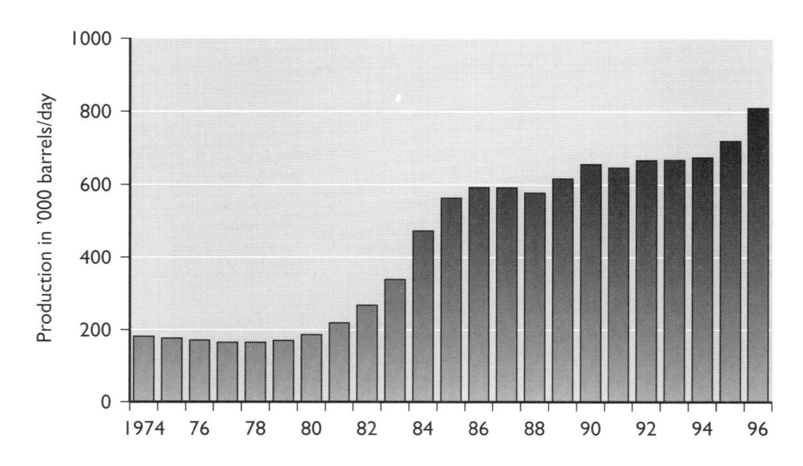

Figure 4.9
Oil production, '000 barrels/day

Oil consumption has increased considerably with the success of the Plano Real economic stability plan, averaging about 1.6 billion barrels a day in 1996. What oil is not supplied by Petrobras, the state-owned oil company is imported from Argentina, Venezuela, the Middle East and Africa.

Figure 4.9 shows that domestic production increased markedly in the 1980s. In 1995 daily production averaged 720 000 barrels, 73 per cent from off-shore wells. The principal producing states are Rio de Janeiro (66.3 per cent), Rio Grande do North (11.8 per cent), and Bahia (8.5 per cent). At present Brazil produces slightly more than half of domestic consumption.

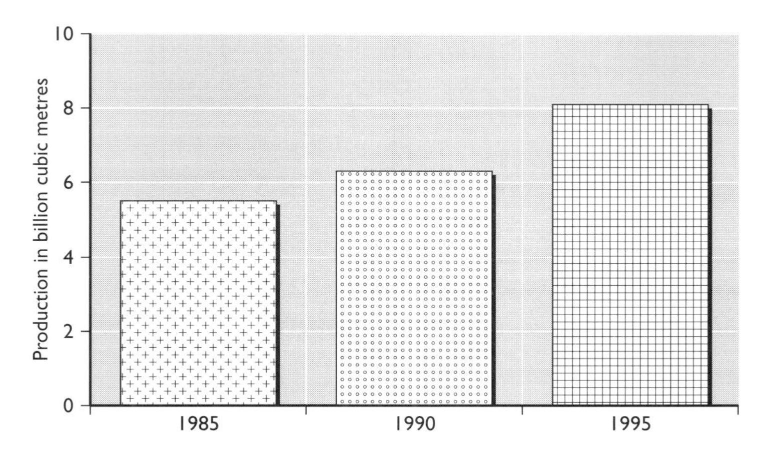

Figure 4.10
Natural gas production, billion m^3

MORE NATURAL GAS

Brazil's natural gas reserves in 1995 totalled 326 billion cubic metres, of which 208 billion were considered proven. This amounts to 0.1 per cent of total world reserves. Figure 4.10 shows the increase in annual production since 1985. In 1995 Brazil produced 0.4 per cent of the world total. Currently, 63.6 per cent of production comes from offshore. The major natural gas states are Rio de Janeiro (39.1 per cent), Bahia (20.1 per cent), and Rio Grande do North (10.4 per cent).

OFFSHORE DEVELOPMENT

Brazil's oil and natural gas industries are dominated by offshore wells, with the greatest concentration of production in the Campos Basin, on the continental shelf off the coast of Rio de Janeiro (Figure 4.11). The more than 30 oil and gas fields in the Campos Basin lie between 55 km and 130 km offshore. The elongated basin, following the general trend of the coastline varies from 100 m to 2000 m in depth. Understandably, the shallowest and most accessible areas of the basin were the first to be developed but in the last decade production has moved into deeper waters, benefiting considerably from expertise developed in the North Sea. The Campos wells contribute 40 per cent to total national gas production, accounting for 62 per cent of offshore production. Petrobras, Brazil's state oil firm, is investing heavily to increase oil and natural gas output from the Basin. Much recent investment has also been directed to the Santos Basin (Figure 4.12), off the coast of Santa Catarina state. The objective is to have 10 wells in production by the end of 1997.

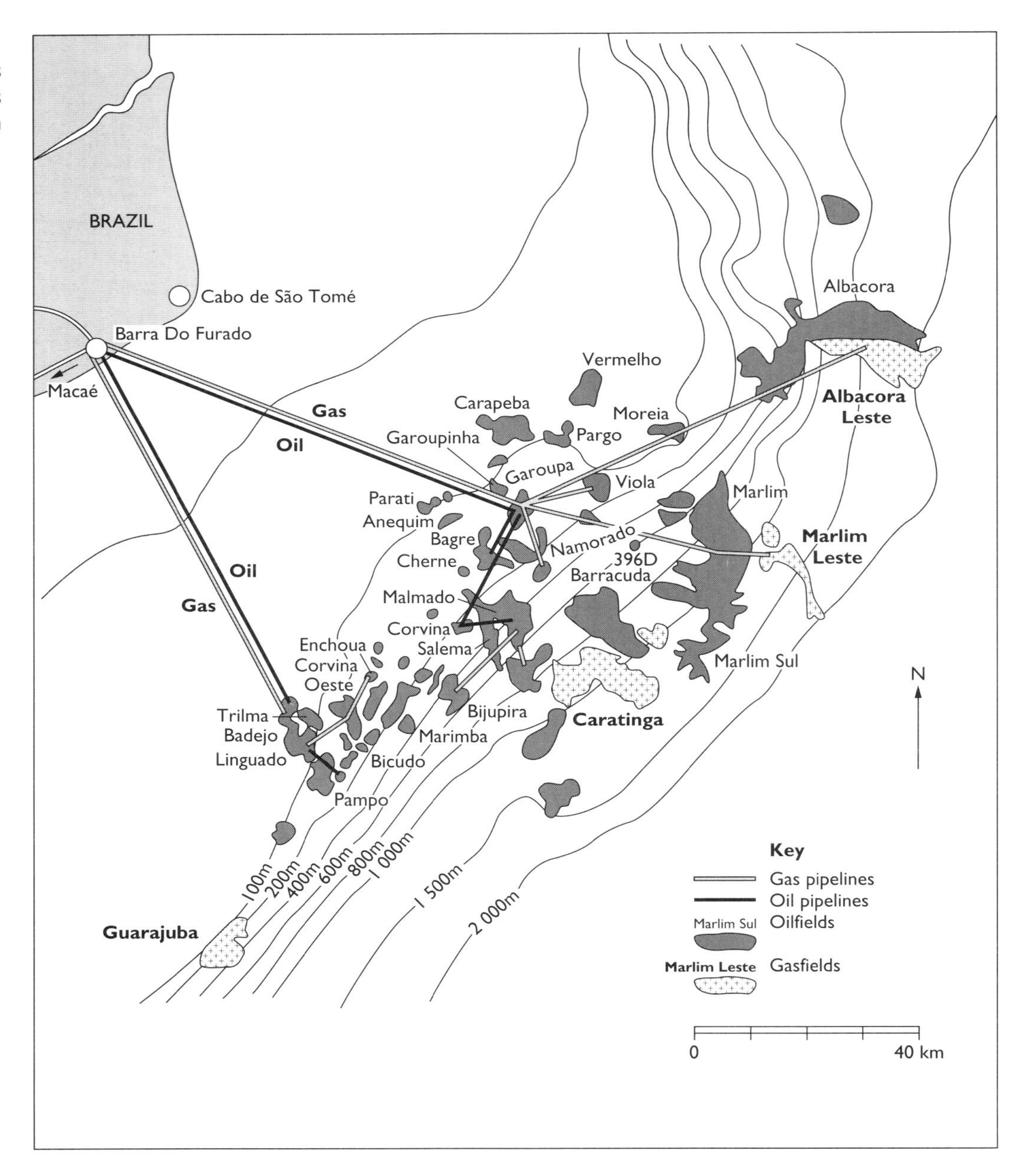

Figure 4.11 Oilfields and gas fields in the Campos Basin

Brazil's state oil firm Petrobras plans to invest $382 million by 1997, the second-biggest sum it has ever spent on a project, to develop a major, shallow-water, oil and natural gas-producing region, the Santos Basin, located off the coast of southern Santa Catarina state.

By 1997, the objective is to produce 30 000 barrels/day and 800 000 cu m/day of natural gas at the Santos Basin, with known reserves of 45.8 million barrels and 3.2-billion cu m of natural gas. The Santos Basin natural gas output doesn't include the volumes now being produced at its already-developed Merluza field, located in the basin.

To reach that production target, Petrobras is spending $59-million for Santos Basin production this year and $323-million in 1996–1997.

That's quite an investment leap from the $17.2-million that Petrobras spent on exploratory investments at the Santos Basin in 1994. Petrobras also is spending an additional $53-million at the Santos Basin this year to do exploratory work in deeper-water fields.

The only place where Petrobras is investing more money is in the already-prolific Campos Basin, by far the country's largest oil-producing region. By 1997, Petrobras will spend well over $1-billion to increase oil and natural gas output at the Campos Basin, a deepwater field just north of the Santos Basin.

In the Santos Basin, Petrobras has already installed one semi-submersible platform connected, via flexible tubing, to one producing well, at a shallow-water depth of 170 metres. That well is producing 8000 barrels/day of oil and 200 000 cu m/day of natural gas. Petrobras also has drilled four more Santos Basin wells at which it is now running long-term production tests, schedules to end in October.

By the end of 1997 Petrobras will have 10 Santos Basin wells in production, serviced by two semi-submersible platforms.

Platt's Petroleum Insight 2, 11 September 1995

Figure 4.12 Developing the Santos and Campos basins

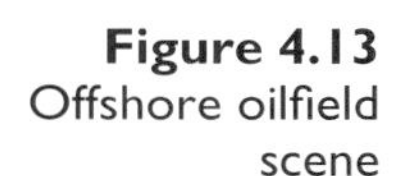

Figure 4.13 Offshore oilfield scene

Figure 4.14 Changes in the Brazilian oil industry

Brazil to end oil monopoly

The lower house of the Brazilian Congress has approved a bill ending the 40-year-old state monopoly of the oil and gas industries – which is expected to lead to private sector investment worth several billion dollars. The bill, approved by 307 votes to 107, will create an independent regulator for the industry, end fuel price subsidies after three years and allow state oil giant Petrobrás to enter into joint ventures with foreign partners. Deputies have still to vote on a controversial amendment allowing for privatisation of Petrobrás.

Geoff Dyer, São Paulo
Financial Times, 14 March 1997

New investment in the Campos and Santos basins should result in a doubling of oil production in Brazil between 1996 and 2000. In the Campos Basin discoveries totalling 1.3 billion barrels were announced in December 1996. Among the new facilities now being installed is the world's deepest offshore well in the Marlim sector, which lies under 1500 m of water. Another new field, Roncador, with a reserve of 1.3 billion barrels, is scheduled to be producing 150 000 barrels per day by 1999.

As these projects come on stream Brazil will achieve the status of a major oil producer and may eventually reduce its import requirement to zero. The ending of the 40-year-old oil monopoly is seen as an important step in this process (Figure 4.17). The 1600 m deep offshore oil wells tapped by Shell Offshore and FMC in the Gulf of Mexico set the world record for deep drilling today, but Brazil is likely to regain the record it formerly held soon.

ENVIRONMENTAL CONCERN

As the infrastructure and production of Brazil's offshore oil and gas industry has expanded, environmentalists and other interested groups have become more and more concerned. Prominent among the latter have been the tourists and fishing industries. As yet, no significant environmental accidents have occurred, but the concerned parties point to the large scale environmental damage caused by huge oil spillage in areas such as Alaska and the Middle East.

? ? ? ? ? ? ? QUESTIONS ? ? ? ? ? ? ?

1. Comment on Brazil's contribution to the global production and reserves of oil and natural gas.
2. Why is Brazil so eager to develop its offshore reserves?
3. Suggest why the oil and gas from the Campos and Santos basins is generally classed as 'high cost production'.
4. Why is the ending of the 40-year-old oil monopoly seen as an important step to increasing the level of investment in the industry?

Brazil's National Alcohol Programme

A RESPONSE TO THE ENERGY CRISIS

Brazil is the only country in the world which mass produces exclusively alcohol powered light-weight vehicles. In the early 1990s this represented about 19 per cent of total output, with a national fleet for this type of vehicle of around five million.

The starting point was November 1975, when Brazil's alcohol programme, 'Proalcool', was officially launched. The sudden rise in the price of oil in 1973 had resulted in major financial difficulties for Brazil as a large oil importer. The alcohol programme became an important part of a wider strategy to cut the oil import bill (Figure 4.15). Alcohol, distilled from sugar cane, has saved a significant amount in foreign currency over the last two decades as well as creating jobs on plantations and in distilleries, the latter benefiting the poor Northeast in particular. Environmental attractions were (1) there was no need to add lead to alcohol fuel because alcohol has a higher octane rating than petrol, and (2) exhaust emissions were fewer.

The programme led to the creation of mini-distilleries in most parts of the country. However an early criticism of the programme was that areas previously used to grow corn, cotton, beans, rice, manioc and other crops were being invaded by sugar cane plantations, thus creating food shortages in some areas.

Ninety per cent of the alcohol produced from sugar cane is used as fuel by around five million vehicles, which in 1994 corresponded to 147 000 barrels of petrol per day. In 1994, 343 distilleries were in operation (178 adjoining sugar cane mills and 165 independent), with an installed capacity of 16 billion litres a year. The state of São Paulo alone accounts for 73.4 per cent of total production.

Figure 4.16 Sugar cane alcohol factory making alcohol for use in cars

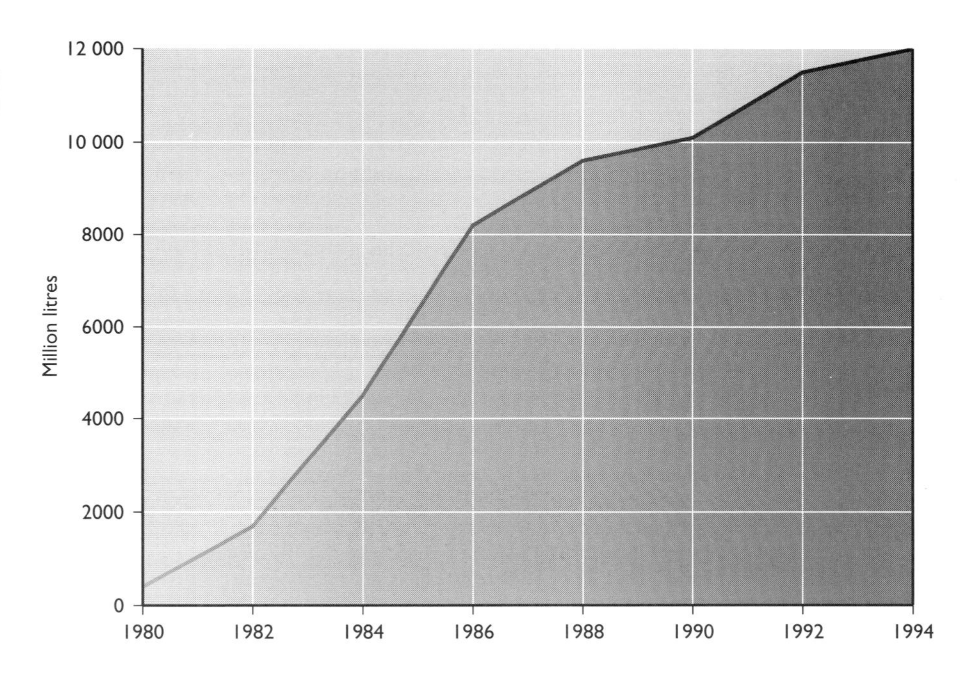

Figure 4.15 Hydrous fuel alcohol consumption in Brazil

WANING POPULARITY

However, the popularity of alcohol cars has waned considerably in recent years as non-alcohol burning car imports have soared. Only 120 000 cars designed for alcohol were sold in 1994 compared with almost 700 000 a year in the mid-1980s. At the height of the programme, in 1985, 85 per cent of automotive industry sales were of alcohol powered vehicles. By 1995 less than 5 per cent of cars in Brazil were alcohol-fuelled. Ironically as Brazilians lose interest in alcohol fuel other countries are assessing its potential. India, where sugar cane production has grown significantly, recently sent a fact-finding team to Brazil to learn more about the industry. For nations such as India which are much further behind in the development spectrum, any measure that limits expensive oil imports merits attention.

BIOMASS: OTHER APPLICATIONS

A number of other advanced forms of biomass energy are either in use or being considered in Brazil. Two examples are:

- In 1993 São Paulo reached an agreement with a group of regional ethanol distillers to increase sugar-powered electricity generation in the state by burning sugar cane pulp in new, advanced gas turbines (Figure 4.17).
- In 1995 Eletrobras announced plans to build a 30 MW biogas-fired combined cycle power plant, in a joint venture with three other interested parties. The plant, estimated to cost a total of $80 million, is being funded by the World Bank, with $23 million coming from the Global Environment Facility. The plant will be fired by gasifying wood chips from fast-growing eucalyptus coppice using technology developed in Sweden. This technology produces for the first time clean enough biogas to be used in a high efficiency gas turbine.

Figure 4.17 Sugar cane as a sensible energy producer

Sao Paulo sugar power

A LANDMARK accord reached by the São Paulo government with a group of 150 ethanol distillers could boost sugar cane-based electricity generation. The state's sugar-power capacity is slated to increase from 200 MW, to 3000 MW, within 15 years.

Investment in the new generating capacity will, if all goes well, be raised by the ethanol producers. The agreement calls for a fifth of the state's electricity to be produced by burning sugar cane pulp. The price is said to be comparable with that of hydro-electric power. And the initiative will save the estimated $8bn capital cost of building a proposed hydro-electric facility of similar capacity.

The arrangement has been made possible by dramatic recent advances in the design of gas turbines and the conversion of solid fuels into gases. It could turn the sugar industry into a net producer of cheap, environmentally-benign energy.

A third of Brazil's 13m cars are fuelled by ethanol which is produced from crushed sugar cane. A byproduct of the process is a dry pulp, or bagasse, which is burned to generate electricity.

A moderately efficient sugar cane processing factory, using conventional technology, can generate about 20 kWh of electricity per tonne of cane, roughly enough to run the plant. But the introduction of a combined bagasse gasifier and steam-injected gas turbine can increase that energy yield 23 times, to 460 kWh/t.

Researchers at Princeton University's Center for Energy and Environmental Studies estimate that the new technologies could increase the potential power capacity from bagasse in 70 sugar-producing developing countries to about 50 000 MW, more than a quarter of their current utility-generated electricity capacity. Bagasse-fired electricity could amount to as much as these countries' entire oil-fired electricity yield.

A world leader in sugar production, Brazil could benefit immensely from the bagasse-based electricity generating process. The São Paulo accord should see environmentally-acceptable power supplied to a rapidly-growing region, while its labour-intensive sugar cane industry should be revitalised, cutting rural unemployment.

An efficient bagasse-based energy technology is especially appropriate for many developing countries. The fuel source is widely distributed throughout rural areas; sugar factories already possess the know-how to produce power; and the simple, modular turbines are easy to maintain.

Petroleum Economist, November 1993

? ? ? ? ? ? ? QUESTIONS ? ? ? ? ? ? ?

1. Why did the Brazilian government initiate the national alcohol programme?
2. Outline the reasons for the declining popularity of alcohol-fuelled cars.
3. What other applications does biomass have to the energy sector?

An Overview

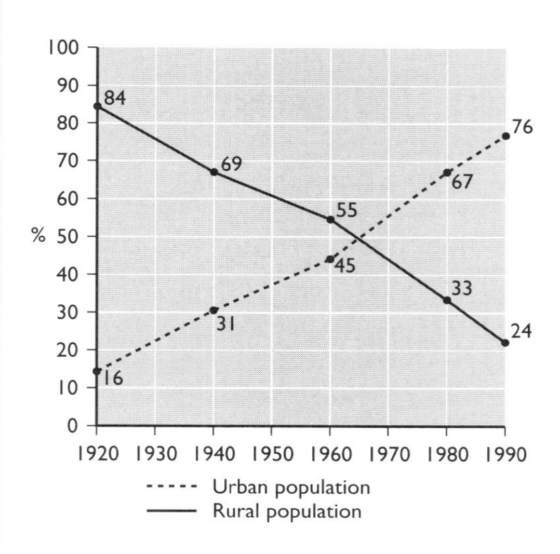

Brazil, like most other countries in South America is highly urbanised. The proportion of people living in cities is far higher than in the other developing continents of Asia and Africa. Figure 5.1 shows the speed at which Brazil has changed from a rural to an urban society. The equilibrium point between rural and urban was reached in the mid-1960s. Between 1970 and 1995 some 30 million Brazilians moved to urban areas, either 'pushed' from their rural environments by adverse factors or 'pulled' by aspirations of a better life in the cities.

Figure 5.1 Urban and rural population change in Brazil, 1920–1990

RURAL PUSH AND URBAN PULL FACTORS

The main push factors responsible for rural to urban migration have been:

- the mechanisation of agriculture which has reduced the demand for farm labour in most parts of the country;
- the amalgamation of farms and estates, particularly by agricultural production companies;
- the generally poor conditions of rural employment. Employers often ignore laws relating to minimum wages and other employee rights;
- desertification in the Northeast and deforestation in the North;
- unemployment and underemployment;
- poor social conditions particularly in terms of housing, health and education.

People were attracted to the urban areas because they perceived life there would provide at least some of the following: a greater variety of employment opportunities; higher wages; a higher standard of accommodation; a better education for their children; improved medical facilities; the conditions of infrastructure often lacking in rural areas; and a wider range of consumer services. The diffusion of information from previous migrants was usually such that most potential new migrants realised fully that there was no guarantee of achieving all or even most of the above but most rationalised that their quality of life would hardly decrease overall. Employment was the key. The most fortunate found jobs in the formal sector. A regular wage then gave some access to the other advantages of urban life. However because the demand for jobs greatly outstripped supply, many could do no better than the uncertainty of the informal sector.

REGIONAL VARIATIONS

Figure 5.2 shows those Brazilian metropolitan regions with a metropolitan population of over one million. Most of Brazil's major cities originally developed around the export of primary products, first in the Northeast and later in the Southeast and South. Later still the rubber boom in the Amazon was reponsible for the considerable increase in size of both Belém and Manaus in the North. In the mid-twentieth century it was industrialisation that really increased the pace of urbanisation with the fastest rates of growth in those urban areas offering the most job opportunities. During this period the pre-eminence of the Southeast was reinforced even further.

Figure 5.2 Populations of Brazil's metropolitan regions with over one million inhabitants

	1980	1991
São Paulo	12 588 725	15 199 423
Rio de Janeiro	8 772 265	9 600 528
Belo Horizonte	2 609 520	3 461 905
Recife	2 347 146	2 859 469
Pôrto Alegre	2 285 167	3 015 960
Salvador	1 766 582	2 472 131
Fortaleza	1 580 066	2 294 524
Curitiba	1 440 626	1 975 615
Brasília	1 176 935	1 596 274
Belém	999 165	1 334 460
Manaus	731 123	1 162 316

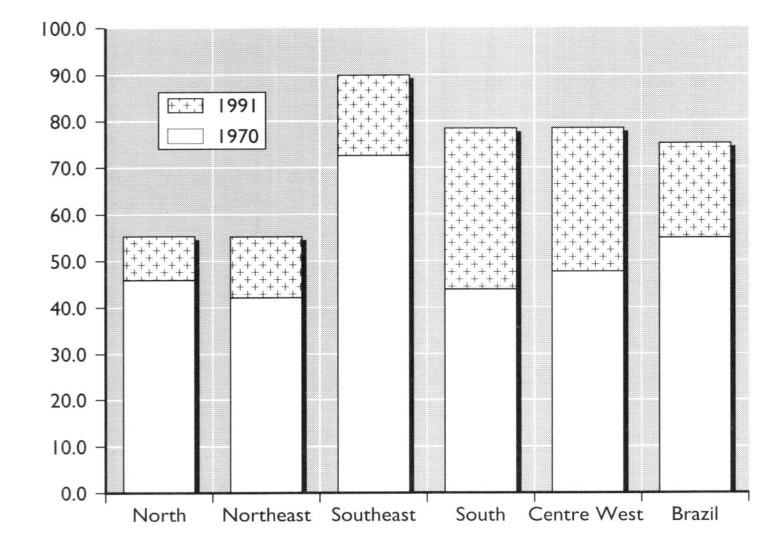

Figure 5.3 Urban populations in the five major regions of Brazil

Urbanisation came last of all to the Centre-West mainly as a result of the decision to build Brasília as the new capital city and other government policies to promote the development of this region. Figure 5.3 shows the percentage of the population classed as urban, by region for 1970 and 1991.

With the exception of Brasília, IBGE has designated the nine largest urban areas as national metropolises (Figure 5.4). Understandably São Paulo and Rio de Janeiro are also considered international metropolises.

Figure 5.4 Populations of the nine largest metropolises

	POPULATION OF METROPOLITAN REGION	POPULATION OF CITY	AREA (KM²)	POPULATION DENSITY* PERSON KM²
São Paulo	15 199 423	9 480 427	7951	1912
Rio de Janeiro	9 600 528	5 336 179	6464	1485
Belo Horizonte	3 461 905	2 048 861	3670	943
Recife	2 859 469	1 290 149	2201	1300
Porto Alegre	3 015 960	1 262 631	5806	520
Salvador	2 472 131	2 056 013	2183	1132
Fortaleza	2 292 524	1 758 334	3483	658
Curitiba	1 975 624	1 290 142	8763	227
Belém	1 334 460	1 246 435	1221	1092

* metropolitan region

THE BEGINNINGS OF DECENTRALISATION

Around the two largest metropolitan areas, São Paulo and Rio de Janeiro, people are dispersing into smaller cities and towns within the hinterlands of these state capitals. In 1991 Greater São Paulo experienced its first negative net migration since the 1940s. This is a trend which is likely to continue and gradually spread to other large metropolitan areas. The large cities continuing to grow at a significant rate are mainly in the North-east. Salvador has very recently passed Belo Horizonte to become the third largest city in Brazil and will continue to attract migrants. Figure 5.5 compares the population growth rates of São Paulo and Rio de Janeiro with other major Latin American cities decade by decade between 1950 and 1990.

Figure 5.5 Major Latin American cities: annual population growth rates, 1950–1990, %

	1950-60	1960-70	1970-80	1980-90
Bogotá	7.2	5.9	3.0	n.a.
Buenos Aires	2.9	2.0	1.6	1.1
Caracas	6.6	4.5	2.0	1.4
Lima	5.0	5.3	3.7	2.8
Mexico City	5.0	5.6	4.2	0.9
Rio de Janeiro	4.0	4.3	2.5	1.0
Santiago	4.0	3.2	2.6	1.7
São Paulo	5.3	6.7	4.4	2.0

FAVELAS AND CORTICOS

Brazil's large urban areas could not cope with the large influx of rural migrants. In the early 1970s around 150 migrants arrived in São Paulo every hour. With no prospect of accommodation in the city itself they put up makeshift shelters (barracos) on the out-skirts of the city. With such a high rate of immigration these makeshift settlements or 'favelas' rapidly expanded in size and number. Figure 5.7 shows the spatial distribution of favelas in Brazil in 1991. In the city of São Paulo in the same year it is estimated that 8.9 per cent of the population lived in favelas. However another 52 per cent were classed as living in 'corticos', overcrowded and decaying buildings in the city itself. The squalid conditions that people have to endure in such areas have been well documented elsewhere.

Figure 5.6 Rio's latest plan to modernise the favelas

The city's latest renovation, devised by its current mayor, Cesar Maia, is called Rio-Cidade, or Rio-City. It includes a $220m scheme to reshape 19 neighbourhoods, and a $600m programme, partly financed by the Inter-American Development Bank, that will bring tarred roads, electricity and sewerage to the *favelas*, the slum-cities that today house a fifth of Rio's people. A 27-kilometre (17-mile) expressway, the 'Yellow Line', is to traverse the city diagonally.

No one could deny that Rio needs work. Over the years, for all its succession of grand plans, the city has changed more than it has been improved. Rio-Cidade sets out to replace century-old sewers, fractured pavements, dilapidated parks and darkened street-lights. The new expressway will—though for how long?—do much to relieve the eternal snarls of traffic. Slum-improvement is desperately needed. Numbers are swelling in the *favelas* twice as fast as in the rest of the city. They are the strongholds and safe houses of its abundant crime.

Yet the current wave of urban reform is everything the opinion-makers love to hate: it is bold, chaotic, autocratic and arbitrary. Not just does the frenzy of public works snarl up the city's already congested traffic. Critics complain that it is merely cosmetic, more a facelift than the real surgery that Rio needs, and does little for public transport, education or health.

The Economist, 28 September 1996

Some favelas are found in close proximity to very affluent areas, a situation known as urban dualism. Rocinha, Brazil's largest favela, with an estimated 200 000–300 000 inhabitants, is in Gavea, one of Rio's richest neighbourhoods.

Efforts have been made to improve the living conditions of the urban poor in various cities at different times. While there have been some notable successes the verdict has often been 'too little, too late'. Rio is the location of one of the most ambitious schemes (Figure 5.6).

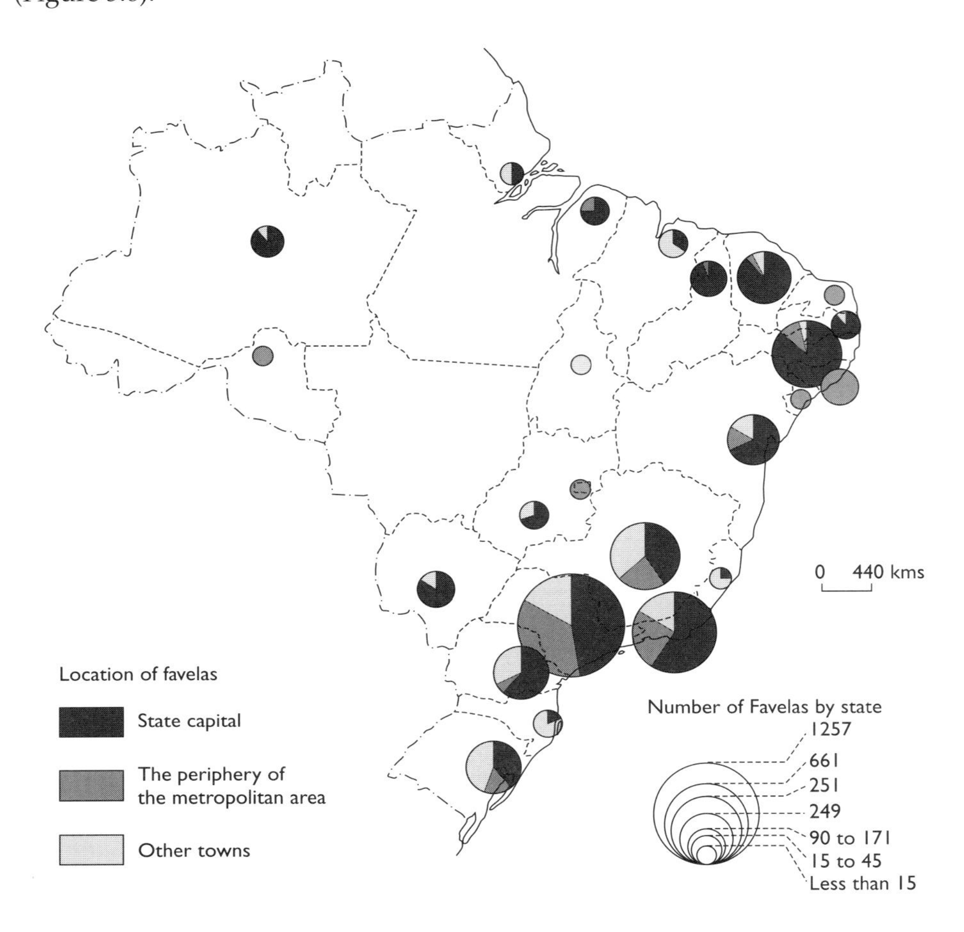

Figure 5.7 Spatial distribution of favelas, 1991

QUESTIONS

1. How do the trends illustrated in Figure 5.1 compare with the UK during the same time period?
2. Plot the information shown in Figure 5.2 on an outline map of Brazil.
3. Suggest reasons for the differences shown in Figure 5.4.
4. Compare the growth of São Paulo and Rio de Janeiro with other large Latin American cities (Figure 5.5) between 1950 and 1990.
5. Analyse the spatial distribution of favelas shown in Figure 5.7.
6. How would you expect living standards to differ between the favelas and the corticos in São Paulo?
7. Comment on Rio de Janeiro's latest attempt at urban renovation.

LOCATION

São Paulo was established as a mission station by Jesuit priests in 1554 near the confluence of the Rio Tietê and a southern tributary, the Tamanduatei. Initially called São Paulo dos Campos de Piratininga, it was located 70 km inland at an altitude of 730 m, on the undulating plateau beyond the Serra do Mar. The cooler and, healthier climate attracted settlers from the coast while the Paraná river system facilitated movement into the interior.

EARLY DEVELOPMENT

In 1681, São Paulo, as the settlement became known, became a seat of regional government and in 1711 it was constituted as a municipality. Coffee was the catalyst for the rapid growth in the latter part of the nineteenth century that transformed the city into a bustling regional centre. It became the focus of roads and railways and its prosperity was assured by a rail link with the port of Santos, completed in 1867. This was the only major routeway scaling the great escarpment of the Serra do Mar. The profits from coffee were invested in industry and by the end of the century São Paulo had become the financial and industrial centre of Brazil. The wealth of the coffee barons was lavished on sumptuous town houses, and prestigious public buildings mushroomed in the business district, the Triângulo.

TWENTIETH CENTURY GROWTH

São Paulo's population growth was relatively slow until the late nineteenth century. In 1874 its population was only about 25 000. However the rapidly increasing demand for labour encouraged immigration and the city's population soared to almost 70 000 by 1890 and reached 239 000 in 1910. The city reached 'millionaire' status in 1934.

By 1950 the population had grown to 2.2 million and São Paulo had clearly established its dominant role in the urbanisation of Brazil. Thereafter the population of both the city and the metropolitan area grew rapidly (Figure 5.8) with the latter reaching 16.5 million in 1995, making it the fifth largest city in the world. The annual growth in population is currently 200 000 in the city and 340 000 in the metropolitan area (Figure 5.9). However, a marked change occurred between 1980 and 1991 (Figure 5.10) with negative net migration recorded for both the City of São Paulo and Greater São Paulo.

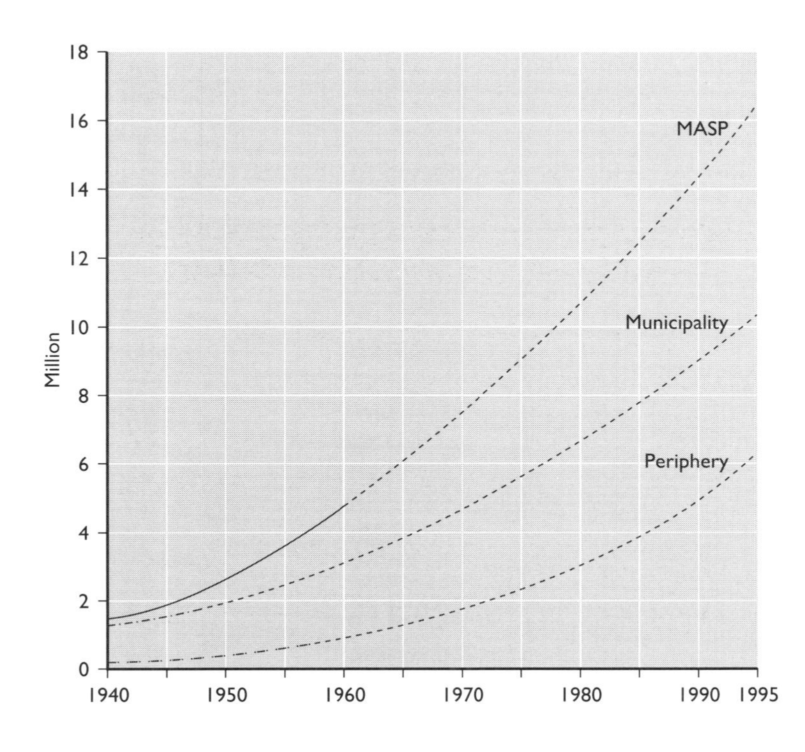

Figure 5.8
Population growth in Greater São Paulo, 1940–95

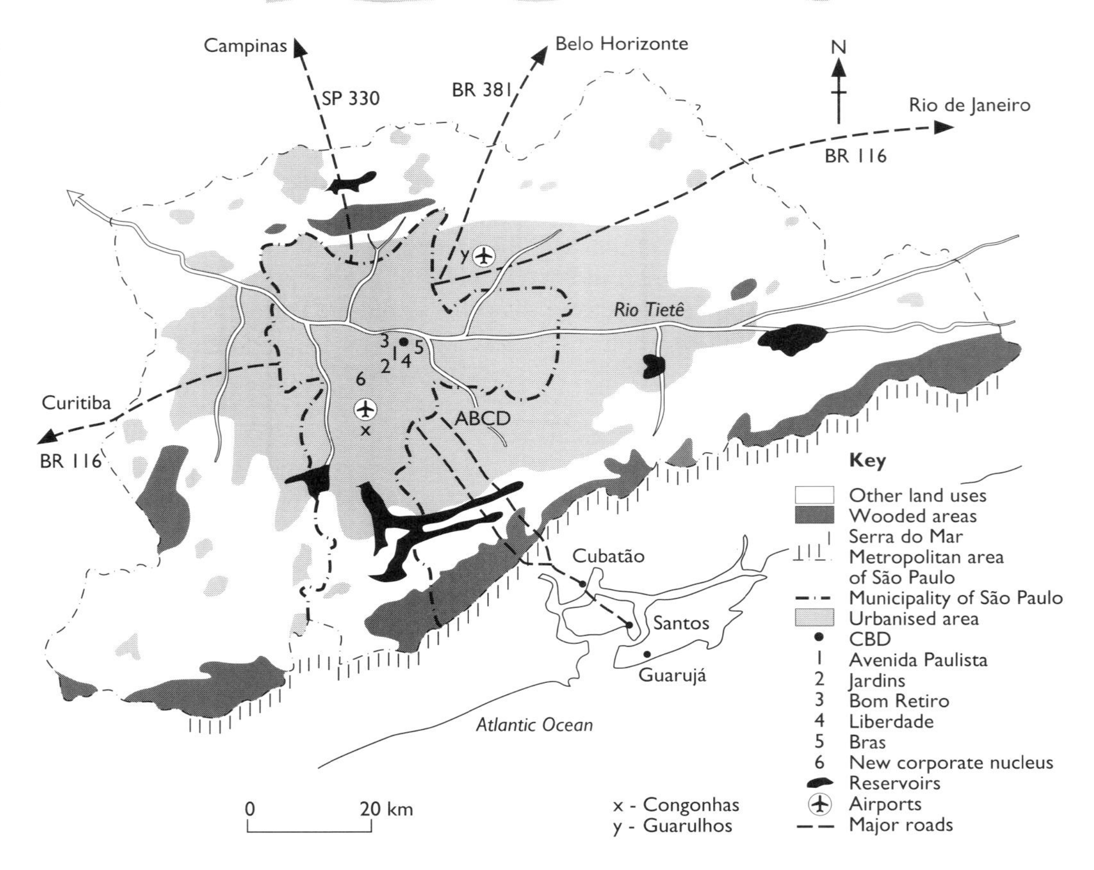

Figure 5.9 Land use in Greater São Paulo

Figure 5.10 Population change: natural change and migration, 1970–91

		GREATER SÃO PAULO	CITY OF SÃO PAULO	STATE OF SÃO PAULO
POPULATION	1970	8 139 730	5 924 615	17 771 948
	1980	12 588 725	8 493 226	25 040 712
	1991	15 416 416	9 626 894	31 546 473
1970/80 CONTRIBUTION TO CHANGE	ABSOLUTE INCREASE	4 448 995	2 568 611	7 268 764
	NATURAL CHANGE (%)	26.45	24.05	6.28
	NET MIGRATION (%)	28.20	19.31	17.35
1980/91	ABSOLUTE INCREASE	2 827 691	1 133 668	6 505 761
	NATURAL CHANGE (%)	24.64	22.25	23.64
	NET MIGRATION (%)	−2.18	−8.90	2.34

A COSMOPOLITAN ENVIRONMENT

In 1934, when the city first reached 'millionaire' status, it was already a rich dough of Portuguese, Italian, African, Spanish, Arab, Jewish, German, Japanese, Armenian and many other immigrants and immigrant descendants. The city houses the largest 'German industrial town' out of the Ruhr and a population of Italian descent second only to Rome. The presence of the Portuguese and Italians is overwhelming in the streets of the metropolis. However, nearly 500 000 Japanese and Japanese descendants, 70 000 Jews, 30 000 Koreans and countless Lebanese, Syrians, Spaniards, Germans, Americans, Scandinavians and so on, do not feel out of place.

The district of Liberdade (Figure 5.14), with its 20 blocks of colourful architecture is the redoubt of the Japanese, whose children account for 13 per cent of the students enrolled in courses at the University of São Paulo. The Jewish population is most concentrated in the neighbourhood of Bom Retiro, one of the most important areas for the garment industry. In the 1970s the development of a Korean settlement started, and today the Koreans own over two thousand garment factories. The 25 de Marco Street is the heart of the Arab community, a 'little Baghdad' of three thousand stores concentrated in less than four kilometres of streets. The neighbourhoods of Bexiga, Brás and Barra Funda are the centres of the Italian community.

ORIGINAL MONO-NUCLEAR FORM

In the early part of the century São Paulo retained its mono-nuclear structure around the Triângulo, its CBD. From here emanated radial highways which were connected by a concentric sub-system which often acted as 'barriers' between different socio-economic areas. In general, residential areas became progressively poorer with distance from the centre, reflecting limited intra-urban mobility and economic concentration in the central area.

METROPOLITANISATION AND INDUSTRIALISATION

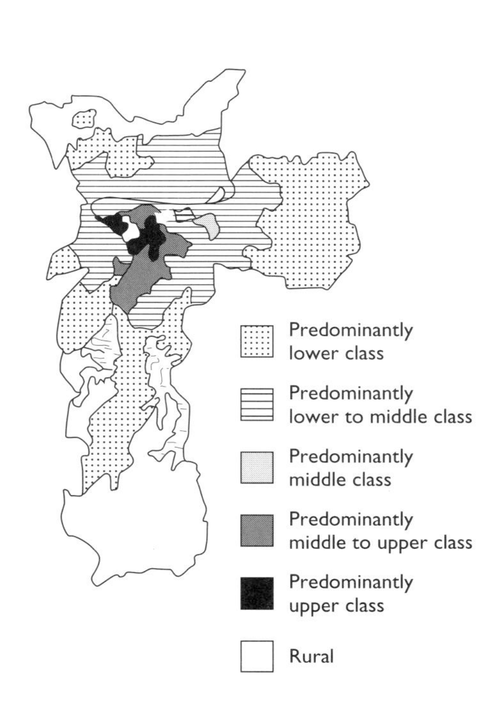

Figure 5.11 Population distribution by class in the São Paulo metropolitan area

The 'metropolitanisation' of the city occurred between 1915 and 1940 due to high rates of both natural increase and immigration. During this period industrialisation had a marked impact on urban structure. The first industrial districts had been located on the south bank floodplain and terraces of the Tietê, but in the 1930s new industrial satellites emerged in the south east. These areas were to grow rapidly in the following decades and now form the most important industrial region in metropolitan São Paulo, known as the 'ABCD' complex after its constituent districts of Santo André, São Bernardo, São Caetano and Diadema. This is the focal point of Brazil's motor vehicle industry and also the location of a wide variety of other industries.

Until the late 1960s the city largely grew unplanned. The Triângulo had developed all the characteristics which make it indistinguishable from any other Central Business District, and the 1970s, particularly, saw a huge increase in the construction of high rise apartments for the growing middle class in the inner area (Figure 5.11).

The ravines of the River Tietê had become sites for numerous favelas (many now removed to make way for inner city parklands). And some of the old mansions near the centre had deteriorated into run-down multi-family dwellings.

Low-grade apartment blocks were built for the workers in the industrial suburbs, but demand far exceeded supply and the authorities were obliged to resort to site-and-service schemes. These too, were insufficient, and large areas of unplanned favelas developed. Meanwhile, the very rich had begun to move out to estates within commuting distance of the Triângulo via the new motorways. Here they live in luxurious villas set in extensive gardens, often with swimming pools and staff quarters. Security men stand guard at the entrance.

Property speculation has resulted in large undeveloped areas within the conurbation, which could accommodate a two-thirds increase in population without any further expansion if they were put to use. According to official data 25 per cent of the city's 1500 km^2 remain unoccupied. However, the forested water catchments of the uplands to the south and to the north are now protected by the 1975 Water Resources Protection Law. However, the two main rivers are said to be dead – most residential sewage and industrial effluent flows directly into rivers and reservoirs; only 10 per cent of the solid waste is collected and treated.

The rapid and generally unplanned growth of São Paulo means that the functional differentiation of different parts of the city is less clearly defined than in most developed world cities. In some cases, squatter settlements are large swathes of urban territory housing hundreds of thousands of people. But more often they are islands of shanty housing, located on vacant land along railway tracks, under flyovers, and in other areas unattractive to the more affluent (Figure 5.12).

Figure 5.12 Location of favelas in the municipality, 1987

	NUMBER	PERCENTAGE
On the banks of streams and rivers	783	49.2
Areas subject to flooding	512	32.2
On steep slopes	466	29.3
In areas of accentuated erosion	385	24.2
On rubbish tips	30	1.9
Beside major roads	40	2.5
Beside railways	25	1.6

FROM HORIZONTAL TO VERTICAL DEVELOPMENT

In the last 30 years São Paulo has lost its characteristic of being a predominantly horizontal city. As competition for land increased it became more important to take full advantage of inner locations and São Paulo embarked on a period of intense vertical development, a phenomenon in Brazil that first appeared in Rio de Janeiro, a city squeezed between the mountains and the sea. A contributary factor to the high rise boom has been the increasing concern over crime which reduced single-family residential security to the point where many affluent homeowners began to opt for well protected apartments.

THE ESTABLISHMENT OF THE METROPOLITAN REGION

It was not until 1968 that the first attempts to formulate a comprehensive planning strategy were made, resulting in the establishment of the metropolitan region in 1973. The fundamental objective was to coordinate administration and planning at the various levels of government and in 1974 the Metropolitan Planning and Administration System began to develop the necessary operational tools. The original mono-nuclear city had developed into a poly-nuclear conurbation, a fundamental fact recognised by the Metropolitan Planning Corporation (EMPLASA) when it set out its objectives.

REDEVELOPMENT OF THE CENTRAL AREA

The Anhangabaú Valley is the historic heart of the city and for decades was the centre of the economic and cultural activities of São Paulo. However the old nucleus gradually lost its significance with increasing suburbanisation and the huge rise in traffic volume which saturated the Valley with over 12 000 vehicles per hour. But in December 1991, the old centre recovered its vitality. After a five year reconstruction, the 77 000 square metre area (one-seventh green space) was reopened to the 1.5 million people who daily cross the centre of the city. Traffic now runs underground in tunnels 570 metres long that connect the northern and southern sections of the city. Much of the area is now pedestrianised and the most important historic buildings have been renovated.

Figure 5.13 Aerial view of the central area

THE JARDINS – INNER CITY AFFLUENCE

No more than a few kilometres southwest of the central area are the most elegant residential areas, comparable to European and North American suburbs. In this area, known as the Jardins (Figure 5.14), there are 50 m^2 of greenery per capita; houses average 60 m^2 per dweller, and the standard of living is that of the upper middle class in most developed countries. The area was laid out in 1915, following the British idea of a garden suburb. It is now dominated by expensive apartment buildings protected by high railings and comprehensive security systems. These exclusive residential neighbourhoods have long since taken over from the city centre as the location of most of the city's best restaurants and shopping streets.

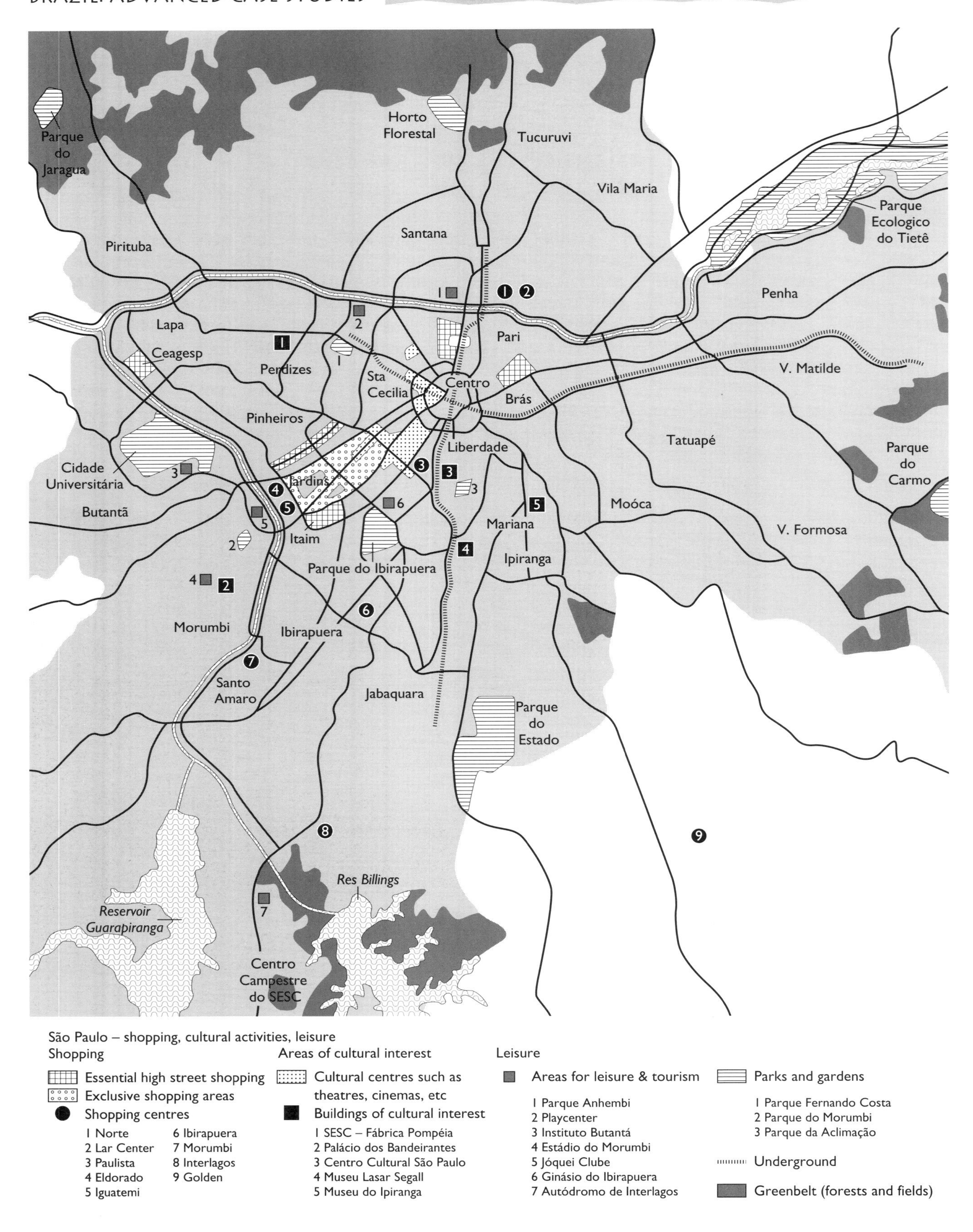

Figure 5.14 São Paulo features and areas of interest

AVENIDA PAULISTA – SÃO PAULO'S FINANCIAL CENTRE

To the southwest of the centre and separating it from the Jardins is the 3 km long Avenida Paulista, the most important financial centre in the city. It is the address of many of the largest banks in the country, two business federations and hundreds of corporations. Mansions once lined this fashionable routeway, but most were replaced in the late 1960s and in the 1970s by skyscrapers. Real estate along the avenue alone is worth $7 billion. The Museu de Arte de São Paulo (MASP) is also located on the avenue.

THE SOUTHWEST – THE NEW CORPORATE NUCLEUS

The corporate addresses in large cities like São Paulo are often related to major road foci. First it was the Anhangabau Valley, then Avenida Paulista and most recently the road alongside the Pinheiros River. A privileged rectangular area served by four of the most important avenues in the city: Juscelino Kubistchek, Morumbi, Luis Carlos Berrini and Nacoes Unidas has become the favoured location for corporate headquarters. It is not only easy access that favours this location. Modern concepts of integrated management require large areas – often over 1000 m^2 per floor, impossible to find in other already crowded commercial areas of the city.

Figure 5.15 Affluent apartment blocks interspersed with low rise development in the Jardins

From 1986 to 1991, nearly 100 large corporations moved their head offices to modern buildings in this area. Residents include Philips, Dow Chemical, Johnson and Johnson, Hoescht, Autolatina, Fuji and Nestlé. Finance conglomerates like Chase Manhattan and Deutsche Bank also have there offices there. The impressive World Trade Centre complex which opened in 1995 acts as the focal point of this office region. Between 1990 and 1993 the area accounted for almost 70 per cent of all new office building in the city.

SHOPPING MALLS

Service industry occupies 54.4 million m^3 of built area, while manufacturing facilities take up 29.5 million m^3. The city has 55 000 shops and 11 shopping malls. The largest shopping mall, Center Norte, is located on the north bank of the River Tietê, about four kilometres north of the CBD. It contains 483 shops on one floor and the parking area can accommodate 17 000 cars. Figure 5.14 shows the location of Centre Norte and the other large shopping centres, the route of the metro and other features of interest close to the centre.

Figure 5.16
A large modern shopping mall

TRANSPORTATION

Figure 5.17
The expanding subway of São Paulo

PUBLIC TRANSPORT

As to transportation, investments will convert to the bus fleet and the subway. São Paulo wants to provide a better and faster transportation to the 3.4 million passengers that use public transportation daily.

In 1991, 75 percent of the users depended on the ten thousand urban buses (57.6 percent on urban lines, and 17.6 percent on inter municipal lines). A little over 16 percent of the passengers used the subway and 8.7 the suburban trains.

Regardless of who is next in the city administration, the trend is to expand the services. The need to improve the rate of 1 bus for each 1,000 inhabitants will also mean an increase in demand for buses from the automotive industry and consequently more jobs.

Most bets, however, are placed on the expansion of the subway network. Today, there are two major lines connecting the North to the South, and the East to the West of the city, and a short branch under Avenida Paulista that will connect Vila Prudente to the bohemian neighbourhood of Vila Madalena. The blueprints for two other major lines are being prepared: one will connect the town centre to the populous neighborhood of Pinheiros, in the West, with a connection to the railway that runs parallel to the Pinheiros River, where new stations will be built, and a branch of the basic North-South line is going to the industrial ABCD, starting at the industrial pole in Diadema.

Government plans to implement a suburban railway compound, to the advantage of the whole South Zone. A large terminal on Florida Street, that will receive subway and railway trains, and a new line of trolley buses, 33 kilometers long.

'Sao Paulo, Brazil,' 1993

It has been estimated that a quarter of all vehicles in Brazil circulate in São Paulo. Car ownership in the city is rising fast and much has been spent on roads to accommodate this trend. Over 35 per cent of households in the municipality own a car. The noise of traffic on the main roads into and out of the city is incessant 24 hours a day. The answer to at least part of the problem is to invest more in public transport (Figure 5.17).

RUBBISH

The City of São Paulo spends $1 million a day on rubbish collection. The cost has risen sharply over the last decade because of (1) a lack of strategic planning; (2) a growing population; and (3) the rising amount of rubbish per person because of increased consumption. Cost is only one aspect of this problem. The other is physical disposal; at present the city has only two landfills for rubbish. In an effort to resolve the latter, two enormous waste incinerators, burning 7500 tonnes a day, are expected to begin operation in São Paulo in 1999.

TACKLING THE FAVELA PROBLEM – THE CINGAPURA PROJECT

This recently announced project has been named after Singapore's huge slum clearance programme. It aims to replace hundreds of favelas with low-rise blocks of flats. The inhabitants – the 'favelados', are removed to army-like barracks while construction proceeds. Residents will pay for their new homes with low interest 20 year mortgages. The Cingapura project is massive: the initial target is to resettle 92 000 families (about 500 000 people) from 243 favelas. Yet how far will it go in a city where more than one-fifth of the inhabitants live in shanty towns. Another question is, that even with subsidy, can the poor afford to pay for their new housing and keep it in repair.

Figure 5.18 A favela close to the CBD

THE FUTURE

The population of São Paulo will continue to grow albeit at a rate below many of the smaller urban areas in the region. Planning will become increasingly important if the economic status of the city and the quality of life of its inhabitants is to be maintained and hopefully enhanced. Figure 5.19 identifies some of the major proposals in the Metropolitan Plan for Greater São Paulo.

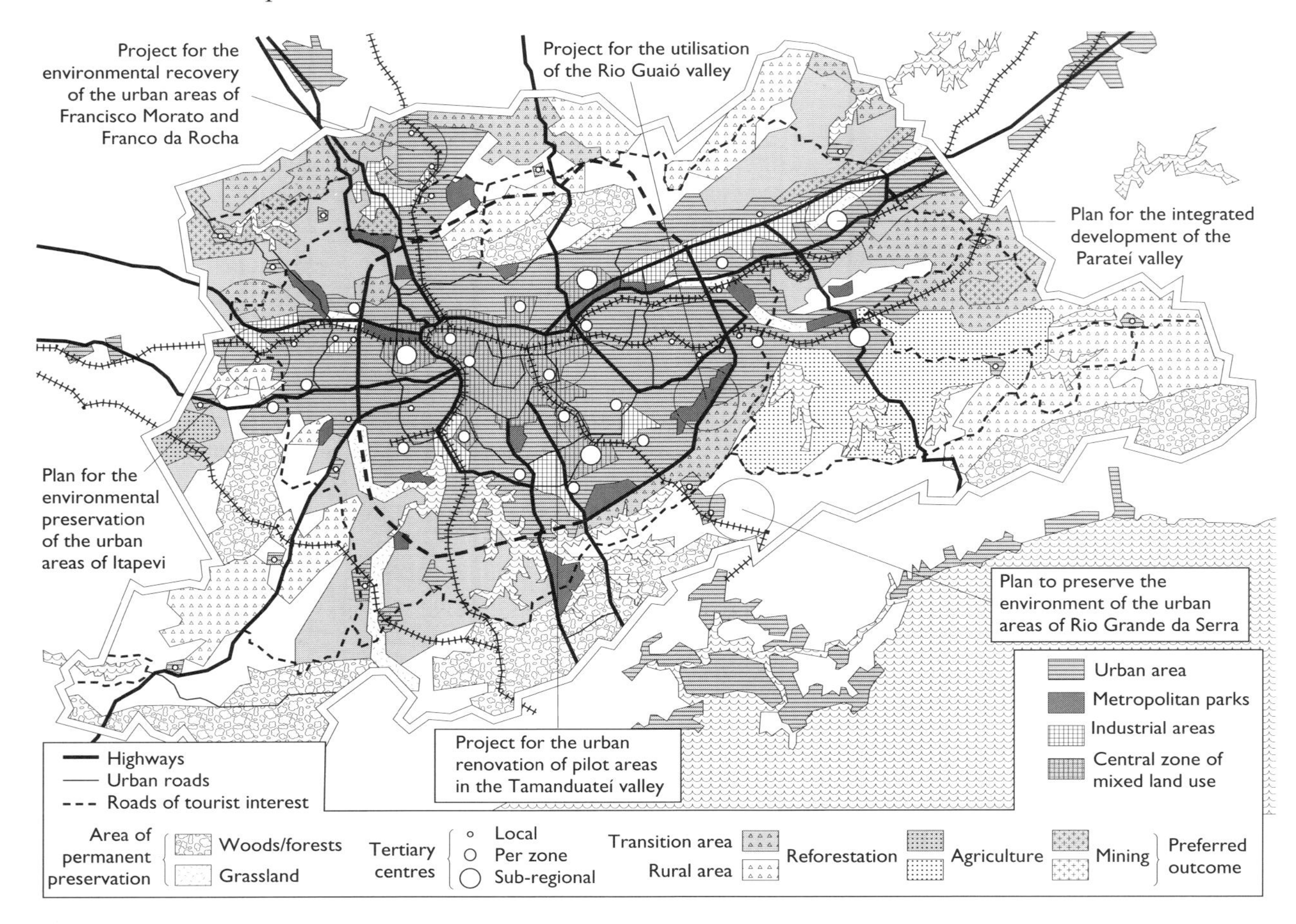

Figure 5.19 Metropolitan plan for Greater São Paulo 1993–2010

? ? ? ? ? ? ? QUESTIONS ? ? ? ? ? ? ?

1. Comment on the rates of population growth shown in Figure 5.8.
2. (a) Describe the location of São Paulo.
   (b) Suggest why this major world city developed on the inland plateau rather than on the coastal strip.
3. (a) Describe the distribution of population in the municipality by social class.
   (b) To what extent and why is this pattern different to European cities?
4. Account for the location of favelas illustrated by Figure 5.12.
5. Compare the Center North shopping centre to the largest shopping centre in your home region.
6. How can the city's transport situation be improved?

Curitiba – Brazil's model city

Curitiba, the capital of Paraná, is located about 80 km inland from the Atlantic coast, and situated on a plateau approximately 900 m above sea level. Its climate is humid sub-tropical with average temperatures ranging from 20.4° C in the summer to 12.7° C in the winter.

Founded as a gold-mining camp in 1693, Curitiba was of little significance until it became the capital of Paraná state in 1853. Since then the city's population has steadily risen: 140 000 in 1940; 500 000 in 1971; and some 1.6 million today. Migration from Europe played an important part in Curitiba's growth. The majority of the immigrants were German, Polish and Italian with smaller groups of French, English, Swiss and Japanese. They settled in small communities in the countryside around the city, most of which were later incorporated into the urban area as Curitiba expanded. Until 1980, Curitiba experienced high rates of population growth, reaching 5 per cent per year. Since then growth rates have decreased to 2.3 per cent per year, reflecting a shift of migratory flows toward neighbouring municipalities. Curitiba used to suffer from most of the problems afflicting other cities in Brazil but recently received the title of 'Ecological Capital' from the United Nations, particularly for the success of its public transport system, its environmental campaign, and its efforts to house and generally improve the quality of life of the poor. Curitiba boasts the highest average standard of living in Brazil with high quality education, health and public transport systems. There are favelas but they are not as extensive as in most other cities and because of the cool, damp winters they are more substantial than those in cities to the north.

Figure 5.20 One of Curitiba's lakeside parks with the CBD in the background. The artificial lakes help to control the floods that once plagued the city by holding excess rainwater and keeping it from inundating low-lying areas.

INTEGRATED PLANNING

Curitiba is generally recognised as the most advanced example of integrated urban planning in South America, combining urban and economic development with considerable improvements in the quality of life of its population. As the architect of comprehensive planning in the city, the former mayor Jaime Lerner states 'Curitiba is different from other Third World cities because it has made an effort to be different'. Beginning in 1970 he launched low-cost programmes to build parks, recycle rubbish, house those on low incomes and develop a mass-transit system. Although the city's planning objectives cover virtually all aspects of urban development it is the mass transit system which is the jewel in the crown.

MASS TRANSIT

The planning process was initiated in 1964 and started to be implemented in 1970. A linear pattern of physical growth along structural corridors was established (Figure 5.21). These structural corridors comprise a three-street system: a central street (a two-way bus lane in the middle adjoined by two one-way slow traffic lanes) and, one block away to either side, high-capacity one-way streets heading into and out of the central area.

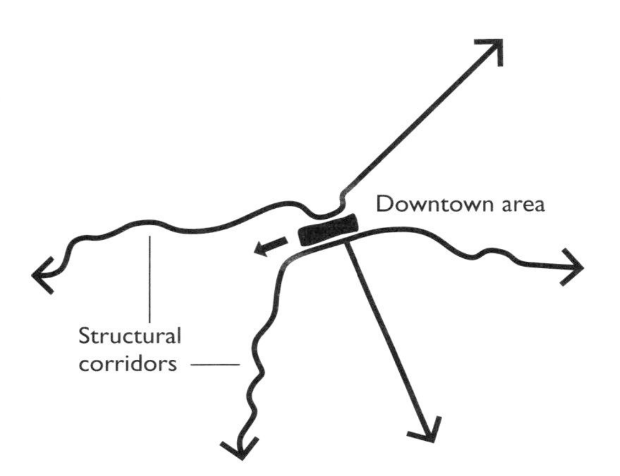

Figure 5.21
Linear growth in Curitiba

The downtown area, where the main streets are pedestrian precincts, is circumscribed by a slow traffic ring. From here Express buses, which use exclusive transit lanes, can be boarded. Bus terminals are located at the end of each Express bus route, allowing for the integration of Feeder buses that bring passengers from the city's outskirts. Another important element in the transportation system is the interdistrict bus system that follows ring routes, connecting neighbourhoods without crossing the central area.

The city's Integrated Transit Network (Figure 5.22) was established in 1980, consolidating the entire bus system through Transfer Stations that allow passengers to transfer to different bus routes for only a single fare. The Network covers about 70 per cent of Curitiba and includes five transit corridors comprising 56 km of exclusive bus lanes. The main corridors are supplemented by 300 km of feeder lines and 185 km of Interdistrict lines. The conventional bus routes plus the Integrated Transit Network serve the entire urban area.

One of the most recent improvements has been the 'Direct Line' system. Direct Line buses have fewer stops and specially designed 'tube stations' to allow faster passenger flow. Fares are paid in advance at the tube station and raised platforms level with the buses' doors allow passenger boarding without any steps. The Direct Line concept was applied to bi-articulated buses, which can carry up to 270 passengers, in 1992. The buses travel through the city at an average speed of 32 km/h. Curitiba has the most efficient transportation system in Brazil with a single fare allowing passengers to ride as many buses as required.

Before the decision to base mass transit on buses, a subway system was considered. However the estimated cost of over $60 million per kilometre was considered prohibitive. In contrast the express bus highways cost $200 000 per kilometre.

On average low income residents spend about 10 per cent of their income on transport, which is below the national average. As a result of the high usage of buses, per capita fuel consumption is 25 per cent lower than in comparable cities in Brazil.

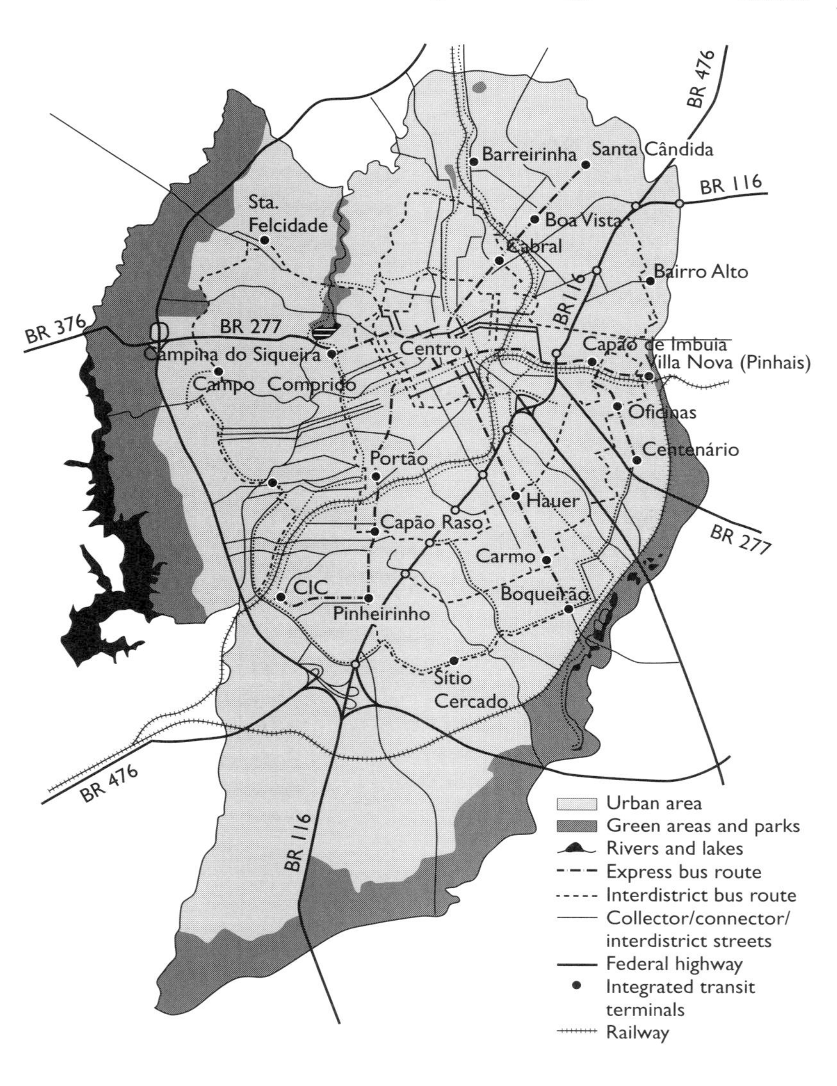

Figure 5.22
Integrated transportation network

Figure 5.23
The mass transit system in operation

THE INDUSTRIAL CITY

During Brazil's 'economic miracle' in the 1970s, Curitiba began to attract large scale industrial investment. The creation of CIC – Curitiba's Industrial City (Figure 5.24), in 1973 was a major part of the objective of upgrading the city's economic profile. Located in a 'green area' about 10 km west and southwest of the city centre and occupying an area of 4370 hectares this major industrial district was planned to minimise adverse impact on the environment. Particular attention was paid to dominant winds, the preservation of flood plains and the protection of water sources in deciding its location.

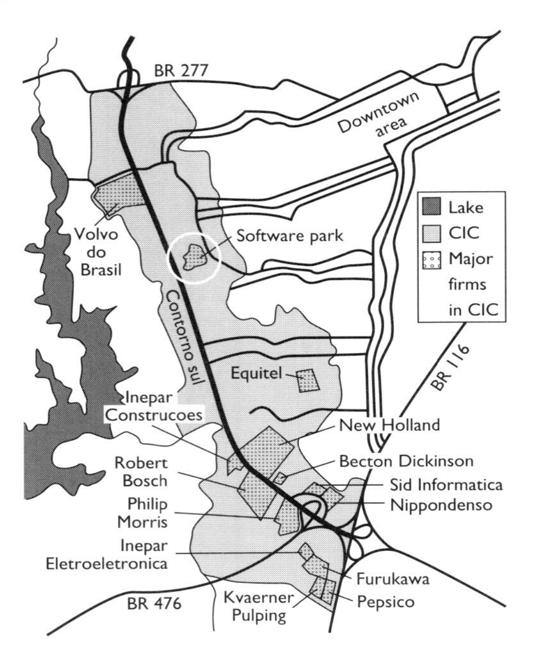

Figure 5.24 Major firms in CIC

The city's planners were convinced that a healthy economy and a healthy environment could not only go hand in hand but that they were of fundamental importance to each other.

A prime planning objective was to integrate industrial activities with other urban functions. Thus housing, leisure areas, public transportation and other services are essential elements of CIC. In mid-1994 there were 16 000 housing units in CIC with another 4500 under construction. Another major objective was the preservation of green areas, aiming not to create an industrial park but a Park with industries. Green areas in CIC total 500 hectares.

Figure 5.25 Curitiba's Industrial City

Figure 5.26 Curitiba's largest export companies, 1993

COMPANY	EXPORT - US$
Phillip Morris	106 277 448
Volvo	98 362 683
Roberto Bosch	88 396 999
Refripar	27 138 082
S.A. Curtume Curitiba	26 560 940
Berneck	17 560 940
New Holland	16 507 344
Selectas	16 745 622
Trutzschler	9 574 021
Compensados Triângulo	7 538 614
LeÆo Júnior	5 929 932
Trombini Embalagens	5 226 564
Total of leading companies	423 510 968
Curitiba	557 773 177
Paraná	2 489 468 902

By mid-1994 500 factories were either operating or in the process of construction, occupying an area of 1255 hectares. These industries generate 50 000 direct and 150 000 indirect jobs. Major companies located in CIC (Figure 5.26) include Volvo (heavy weight trucks), Bosch (injectors for diesel engines), Phillip Morris (cigarettes), and Nippondenso/Toyota (air-conditioners for motor vehicles). The dominant sectors within the Industrial City are; metal-mechanics, electrical and telecommunication products, plastic products and textiles.

In addition to those industries located within the Industrial City, there are elsewhere in Curitiba approximately 5300 industrial enterprises of diverse sizes. With only 16 per cent of the state's population, Curitiba produces 24 per cent of all the goods and services generated by the state. Since the mid-1970s the nature of economic activity in Curitiba has changed significantly (Figure 5.27) and the city now has one of the most positive industry mixes in the country.

Software Park

In an effort to keep up with global industrial development the city has invested in the creation of a Software Park, located in a wooded area within CIC. The Software Park occupies a site of 19 hectares integrated with an environmental preservation area. It aims to provide the physical, institutional and marketing resources to support the implementation and expansion of software engineering firms, and, in order to meet the demand for high-speed digital telecommunications, a fibre optic network linking all firms in the Industrial City is being installed. The idea behind the Software Park is that the productivity of each company located here will be enhanced by links established with other firms in the Park while the utilisation of a common infrastructure will reduce collective costs. Companies also benefit from:

- low land prices for sites;
- institutional support from municipal authorities;
- availability of technical support; and
- access to the laboratories of CITS (International Centre of Software Technology) as well as the local universities.

Curitiba is one of Brazil's 13 Software Export Nuclei, which are part of the National Software Export Program, Softex-2000. The program's goal is to transform this sector into one of Brazil's main exports by the end of the decade.

Figure 5.27 Change in types of industry in Curitiba from 1974–92, %

INDUSTRY	1974	1992
GROUP I – Non-durable consumer goods		
Real Estate	9.48	1.13
Pharmaceutical products	0.14	0.33
Perfumes, soaps and candles	0.96	0.43
Textiles	0.80	0.40
Garment and shoes	1.24	0.60
Food products	11.12	5.80
Beverages	1.94	6.84
Tobacco	0.00	7.52
Publishing and printing	4.32	1.69
Total: Non-durable consumer goods	**30.00**	**24.74**
GROUP II – Intermediary goods		
Non-metallic minerals	2.73	2.04
Metallurgy	9.44	3.52
Timber	24.89	6.69
Paper and cardboard	4.76	2.84
Rubber	1.23	0.13
Leathers and furs	1.84	0.65
Chemicals	7.79	4.94
Plastic products	4.35	3.02
Total: Intermediary goods	**57.03**	**23.83**
GROUP III – Capital and durable consumer goods		
Mechanics	6.95	11.63
Electric and communication material	2.79	9.25
Transportation material	2.16	27.89
Miscellaneous	1.06	2.65
Total: Capital and durable consumer goods	**12.96**	**51.42**

Source: PARANÁ, 1988; 1991; 1994

Other Aspects of Physical and Social planning.

- Total health care is provided for all children up to five years old with health care posts operating twenty-four hours a day.
- Potable tap water is supplied to more than 95 per cent of the population. By 1995, it is expected that virtually all of Curitiba's population will be provided with treated water and, as a result of the Environmental Cleanup Program, more than 85 per cent will have access to sewage collection and treatment services.

- Curitiba has a well developed education system, with compulsory schooling for children between the ages of seven and fourteen. Environmental understanding is an important part of the curriculum and in 1991 the City created the Open University for the Environment. The city has two international schools. There is also a system of libraries, one in each neighbourhood, known as the 'Beacons of Knowledge'.
- In 1970, Curitiba had 0.46 m^2 of open space per person; today its 21 parks and other open spaces provide 52 m^2 per person. Most parks are centred around artificial lakes which were constructed to control the persistent flooding that plagued Curitiba in the 1950s and 1960s. The flooding was caused by urban growth, particularly along the banks of streams and rivers which increased the rate of runoff. Buses integrate the parks with the rest of the urban system.
- Curitiba was the first city in Brazil to recycle its rubbish. The 'Green Swap programme' allows recyclable rubbish to be exchanged for a wide variety of products including fruit and vegetables, transportation vouchers and textbooks. Recycling centres employ recovering alcoholics.
- There are 40 centres that feed street children and teach them simple skills.
- A 100 km system of bicycle paths, the largest network of its type in the country, connects to the bus network and links the city's main parks.

Figure 5.28 Basic development and quality of life indicators for Curitiba

	CURITIBA
Per Capita GDP (US$)	5149
Employed population with income above 2 minimum wages (%)	68.4
Employees officially contributing to social security (%)	72.2
Urban housing units with running water (%)	97.2
Urban housing units with installed electric power (%)	99.0
Urban housing units with sewage facilities (%)	61.0
Urban housing units benefiting from garbage collection (%)	97.5
Privately owned urban housing units (%)	74.3
Infant mortality (per 1000 born alive)	20.3
Basic inoculation coverage (%)	88.0
Telephone terminals (per 1000 inhabitants)	208
Cars (per 1000 inhabitants)	267
Refrigerators (per 1000 inhabitants)	261
Weekly newspaper printing (per 1000 inhabitants)	850
Literacy rate	93.8%
Green area (per inhabitant)	54 m^2

? ? ? ? ? ? ? QUESTIONS ? ? ? ? ? ? ?

1. Suggest why city planners saw economic growth and environmental improvement as fundamentally important to each other.
2. Why did Curitiba opt for a mass transit road system rather than an underground rail system?
3. Analyse the changes in economic activity in the city between 1974 and 1992.
4. Consider all the aspects of integrated planning in Curitiba. Put them in order of importance and justify your ranking.
5. Study Figure 5.28. Comment on the quality of life in Curitiba, comparing it with Brazil as a whole and with the UK.

The background

DIVERSIFICATION BEGINS IN THE 1930S

Figure 6.1 provides a brief review of the history of manufacturing industry in Brazil. While the country clearly recorded industrial growth prior to the 1930s it was not until that decade that industrialisation in the true sense arrived in Brazil. Before this time the range of industrial production was extremely restricted. For example the 1920 Census showed that 47 per cent of the industrial labour force was employed in just two industries; food processing and textiles. At the time agriculture still employed more than two-thirds of the total labour force.

Before the First World War Guanabara, then the Federal district, had more manufacturing establishments than any other state, followed by São Paulo and Rio Grande do Sul. However, by 1919 the state of São Paulo had moved into first place with 31 per cent of all manufacturing firms in Brazil, a ranking it has held ever since. Figure 6.2 shows how the value of industrial production has changed by leading state since 1907.

The shortage of imported goods during the Second World War gave added impetus to the government's policy of import substitution. Between 1947–61 the volume of industrial production tripled and in 1958 Brazil displaced Argentina as the leading industrial nation in Latin America. The car industry was promoted as the key industry around which a range of ancilliary industries would develop.

Manufacturing Industry in Brazil: A Brief History

1. Very limited industrialisation prior to late 1800s – (a) until independence in 1822 Portuguese mercantilist policies prohibited industry in the colony (b) the free trade policy and heavy reliance on raw material exports during the years of the Brazilian Empire 1822–89 presented unfavourable conditions for industrial development.
2. Late 1800s/early 1900s – attempts to achieve economic independence [after becoming a Republic in 1889] spurred industrial growth. Profits from coffee invested in food processing, textiles, iron and steel, and HEP.
3. Between the early 1930s and late 1960s the dominant process was import-substitution industrialisation [ISI]. In the 1930s the emphasis was on the production of non-durable consumer goods.
4. Early/mid-1940s – the government moved from being just a regulator to a supplier of goods and services, exemplified by the construction of the first integrated steel mill by the state in Volta Redonda.
5. Late 1940s/early 1950s – ISI reached a second stage with the expansion of the durable goods sector.
6. Second half of 1950s – government expenditure and foreign capital helped to raise investment to unprecedented levels. Huge public investment in energy and transport; the construction of Brasilia; the establishment of the car industry.
7. Early 1960s – a range of economic problems restricted industrial development.
8. The change to a military government in 1964 brought about economic stability leading to the ''economic miracle'' of 1968–73 when the output of manufacturing industry grew by an average of over 13 per cent per year led by cars, iron ore, steel and petrochemicals. Brazil entered the 1970s with the world's highest growth rate.
9. Mid/late 1970s – after the first oil crisis in 1973/4 growth was achieved in capital goods and basic intermediate goods at the expense of a sharp rise in external debt. Second oil price rise in 1979 created more problems.
10. Recession, inflation and debt in the 1980s and early 1990s provided a very uncertain environment for industry.
11. Mid/late 1990s – industrial expansion led by the car industry with new era of economic stability due to introduction of the Real Plan in 1994. Increasing diversification.

Figure 6.1

The government successfully attracted many large foreign multinationals in this and other sectors of the economy. It is thus not surprising that the downturn marked by the second major rise in oil prices in 1979 came as something of a jolt.

Figure 6.2 Industrial production by state from 1907 to 1990

	% OF TOTAL				
	1907	1919	1939	1970	1990
Pernambuco	7.4	6.8	4.8	2.1	2.1
Bahia	3.4	2.8	1.4	1.6	3.8
Minas Gerais	4.4	5.6	6.5	7.1	8.3
Rio de Janeiro	7.6	7.4	5.0	15.5	8.4
Guanabara	30.2	20.8	17.0	–	–
São Paulo	15.9	31.5	45.4	57.2	51.4
Paraná	4.5	3.2	2.2	4.5	4.3
Rio Grande do Sul	13.5	11.1	9.8	6.3	8.1
Santa Catarina	1.9	1.9	1.8	3.2	3.9
Others	11.2	8.9	6.1	2.5	9.7

THE PROBLEM YEARS OF THE 1980S AND EARLY 1990S

For much of this century manufacturing growth has surpassed overall GDP growth but this was not always the case during this very variable period. This was a problem period for industry because:

- as a large importer of oil the substantial price increase affected all industries using oil either as a fuel or a raw material. The same factor increased the price of almost all imported goods and machinery;
- the increase in global interest rates increased the cost of borrowing for firms and made them less likely to invest in new technology and production;
- the impact of the recession in the developed world led to a fall in international demand for Brazilian products;
- because of the world recession a number of multinational companies reduced their production in Brazil;
- the war between Britain and Argentina over the Falklands Islands made some multinational companies and banks wary of investing in South America for a number of years.

Other developing countries, such as Mexico and Nigeria, which had based their industrialisation on heavy borrowing and investment from multinational companies encountered similar problems.

CURRENT SECTORAL AND SPATIAL PATTERNS

Figure 6.3 provides a breakdown of employment and the value of production by type of industry while Figure 6.4 presents a more detailed spatial analysis. The latter clearly shows the dominance of the Southeast in general and of the state of São Paulo in particular. There is close competition for second place between two more states from the Southeast, Rio de Janeiro and Minas Gerais and the state of Rio Grande do Sul in the South.

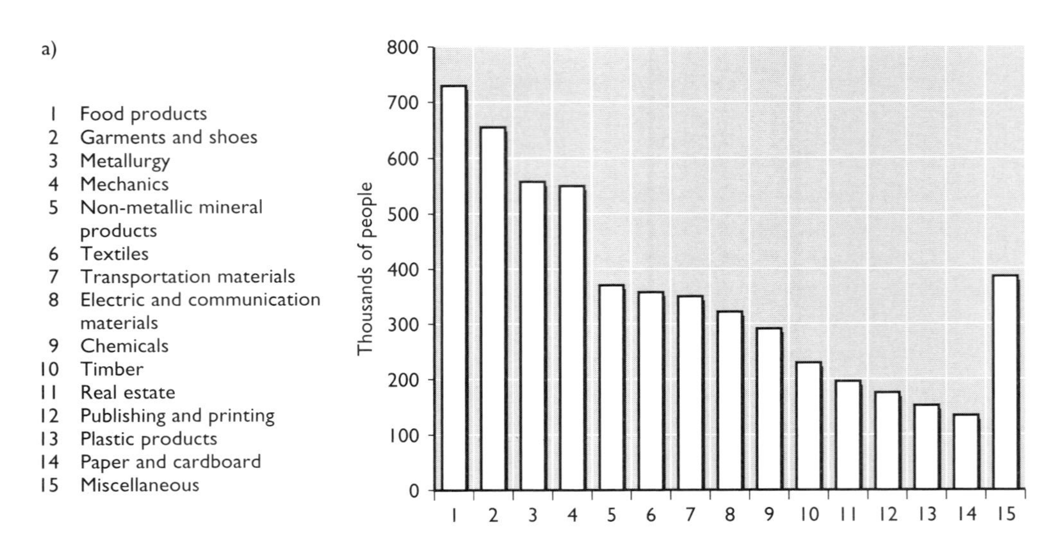

Figure 6.3a Manufacturing industries: employment

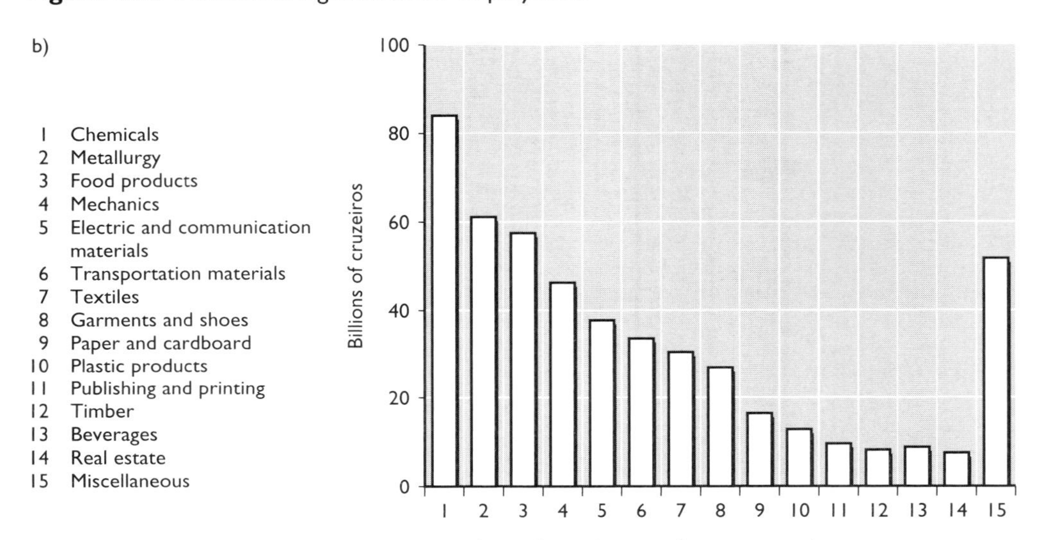

Figure 6.3b Manufacturing industries: value of production (in cruzeiros)

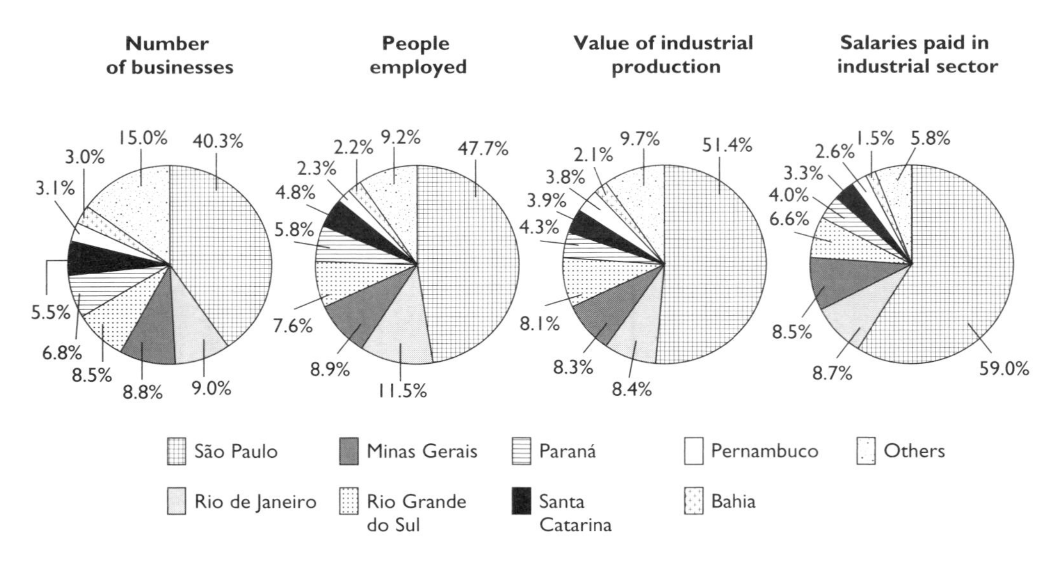

Figure 6.4 Distribution of manufacturing industry within Brazil

FOREIGN INVESTMENT

Investment from abroad remains an essential element of industrial development in Brazil. Figure 6.5 shows that the USA is by far the largest source of inward investment with a total investment stock of almost $36 billion in 1995. Germany was in second place with £11 billion. Although not all of this is in manufacturing industry, the latter remains the main focus of foreign interest when it comes to investment. The successes of the Real Plan in creating a stable economic environment are clear to see in Figure 6.6.

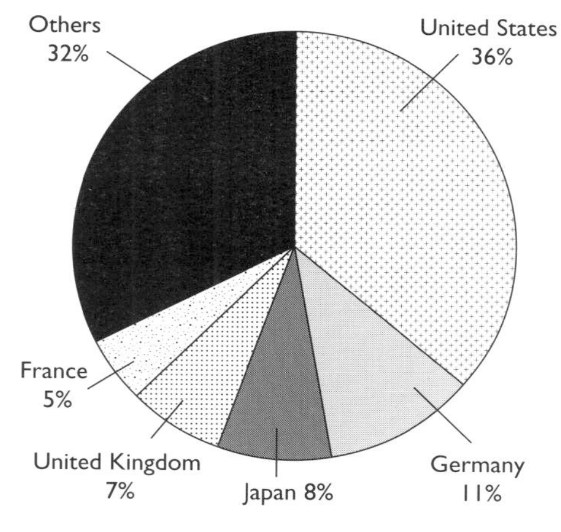

Figure 6.5 Breakdown of foreign investment in Brazil

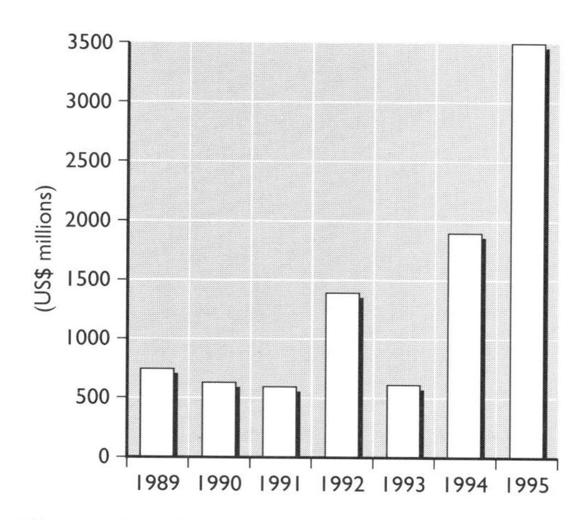

Figure 6.6 Foreign investment 1989–95

STEEL

Steel production is a measure often used to compare the industrial strength of countries.

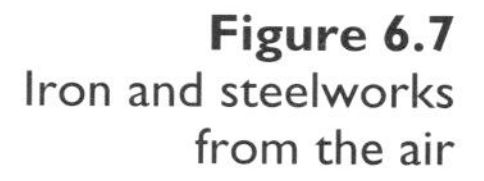

Figure 6.7 Iron and steelworks from the air

Figure 6.8 Steel production, 1960–95

YEAR	MILLION TONNES
1960	2
1965	3
1970	5
1975	8
1980	15
1985	20
1990	21
1995	25

Current production is more than twelve times that of 1960 (Figure 6.8). Brazil is seventh in the world ranking, following China, Japan, the USA, Russia, Germany, and South Korea. Brazil has 21 steel plants in 10 states. However, the largest ranked only 41st in the 1995 world league table. Steel was one of the first industries to be privatised with the switch to private ownership starting in 1989. The fall in inflation has seen the demand for goods such as cars and electric appliances (big users of steel) soar.

However Brazilians still only use 60 kg of steel per head compared to over 300 kg in the USA. In 1994 steel production costs were only 60 per cent of those in the USA. Although Brazilian steel workers were still only half as productive as their counterparts in the USA, they were the most productive in Latin America. The industry in Brazil benefits from cheap electricity and from plentiful supplies of local iron ore.

Figure 6.9 Non-wage labour costs as a percentage of wage costs, 1995

	%
Brazil	81.9
Germany	60
Britain	58.8
Holland	51

LABOUR COSTS

Although wages in Brazil are well below the level of the developed nations, non-wage costs are high (Figure 6.9), a factor which does not help Brazil's competitiveness with other countries at a similar stage of development. The government is currently working on measures to relax the requirements generating these labour-related costs.

Figure 6.10
A manufacturing scene in the informal sector

Such costs form part of what is generally known in the country as the 'Brazil Cost', an accumulation of factors that make the cost of operating in Brazil higher than in competing countries.

THE FORMAL AND INFORMAL SECTORS

A recent study has estimated that half the country's workforce is employed, to a greater or lesser extent, in the black economy, or informal sector. This part of the economy is reckoned to have a turnover of $220 billion, about 35 per cent of the formal economy. The government is making efforts to bring the informal sector into the mainstream economy. For example, two unions for workers in the informal sector are currently being set up.

? ? ? ? ? ? ? QUESTIONS ? ? ? ? ? ? ?

1. Discuss the assertion that 'the process of industrialisation did not begin in Brazil until the 1930s'.
2. What is import-substitution industrialisation? Why did the government foster this strategy?
3. Use an appropriate cartographic method to illustrate the data presented in Figure 6.2.
4. How dominant is the Southeast in industrial terms?
5. Comment on the relationship between the two graphs in Figure 6.3.
6. Why does the government want to bring as much of the informal sector as possible into the mainstream economy?

The Southeast: The Industrial Heartland of Brazil

Manufacturing industry in Brazil is heavily concentrated in the Southeast which is comprised of the four states of São Paulo, Rio de Janeiro, Minas Gerais and Espirito Santo. A further level of concentration can also be recognised because within the Southeast the state of São Paulo dominates the industrial scene, accounting in 1995 for 38.5 per cent of the total tax charged on sales of manufactures in Brazil. About half of all the manufacturing concerns in the country are members of the state's industrial federation, FIESP, which lobbies hard in Congress to protect its members interests. The state is responsible for nearly one-half of Brazil's total GDP. The literature on the subject also refers to the 'industrial triangle' between the region's three largest metropolitan areas; São Paulo, Rio de Janeiro and Belo Horizonte.

The reasons for such a high level of industrial concentration are multivariate. No other part of the country or the continent as a whole possessed such a combination of advantages for the development and expansion of industrial enterprise.

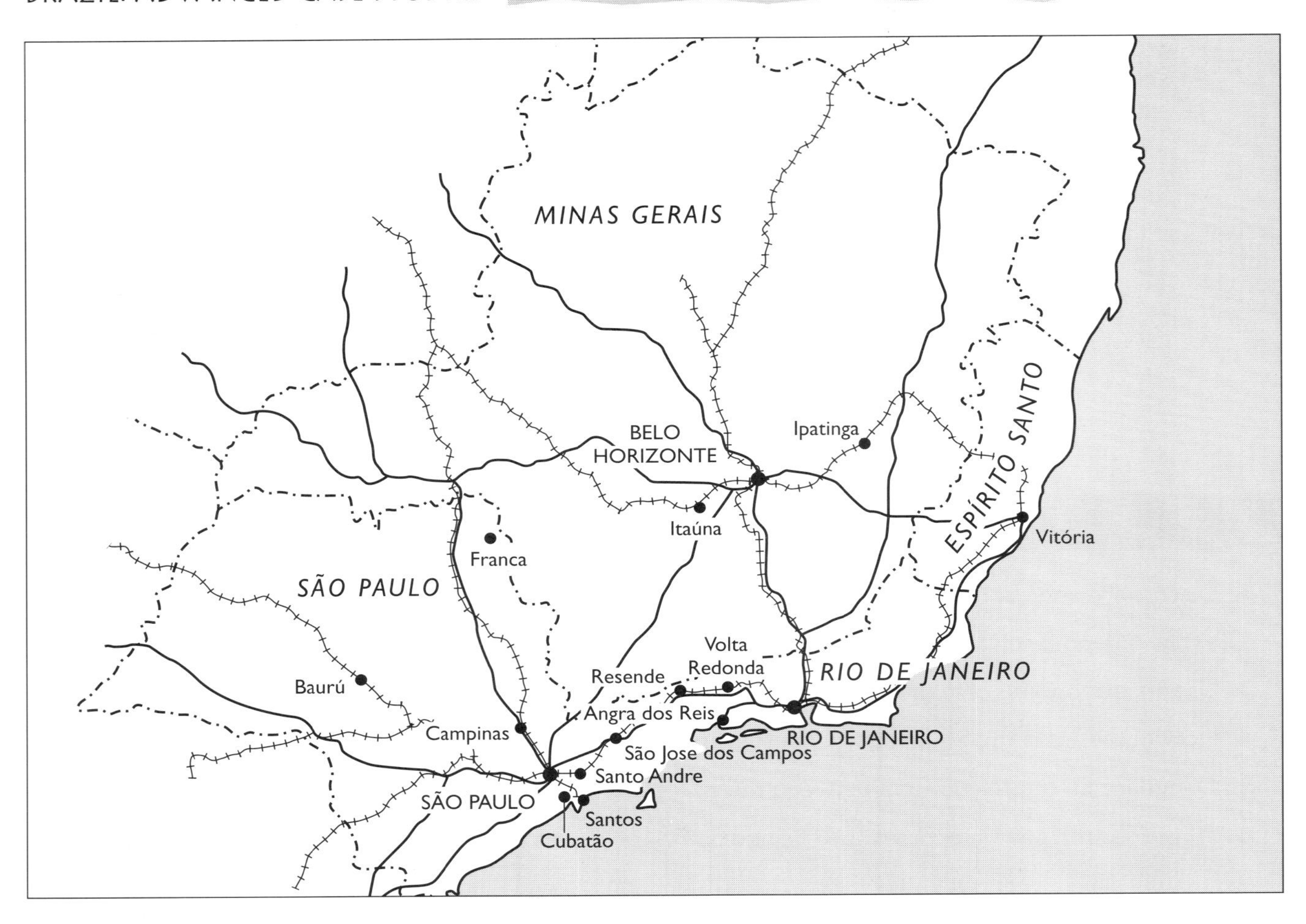

Figure 6.11
Map of the Southeast

THE SÃO PAULO-CUBÃTAO-SANTOS CORRIDOR

São Paulo: Brazil's Major Industrial Centre

Brazilian production of consumer goods began in São Paulo and it was in this city that Brazil took its first steps at domestic production as a substitute for imported goods. São Paulo imports and exports products that account for a third of Brazil's entire trade balance. More than half of Brazil's foreign sales come from the state of São Paulo and 35 per cent from the metropolitan area. A large variety of manufactured products dominate the export scene today, far surpassing the agricultural products that were the export mainstay until the 1960s.

São Paulo is the main centre for both traditional and high technology industry. The former include industries such as food processing, shoes, brewing, furniture, textiles, clothing and printing. The motor vehicle industry was the main catalyst for growth in the post-war period. Ford, General Motors and Volkswagen selected São Paulo as their South American location, attracting hundreds of component suppliers. Other major industries to develop significantly in the metropolitan area were chemicals, iron and steel, cement, heavy and light engineering, pharmaceuticals and plastics.

The city's first industrial districts were sited on the south bank floodplain and terraces of the River Tietê and over the years industry has decentralised, mainly to the south although the main arterial roads to the north, west, and east have also seen significant industrial expansion. Today large companies can be found within the City of São Paulo, either in traditional neighbourhoods such as Ipiranga and Lapa or in the newer south zone. Then, of course, there are the industrial districts in the wider metropolitan area – Osasco, Barueri and the ABCD complex named after the municipalities of Santo Andre, São Bernardo do Campo, São Caetano and Diadema.

Figure 6.12
Made in Brazil

Foreign relations

There is a saying in Brazil that São Paulo is Germany's biggest industrial city. It is a reference to the growing number of German companies operating there. According to Werner Ross, president of the Brazilian-German Chamber of Commerce, Germany has $10 billion invested in Brazil – 15 per cent of the total foreign investment. The close commercial ties between the two countries has a long history; the chamber itself is 80 years old this year. There are now 1,024 German companies in Brazil, mainly in the state of São Paulo, the country's financial and industrial powerhouse.

But few of the 400,000 people of German origin that live in the state work for German businesses. And, increasingly, German executives sent over from the parent company are being replaced by Brazilians. Ross himself, who is president of Degussa, a precious metals and chemicals company in São Paulo, has lived in Brazil for 24 years. When he retires, a Brazilian will take over.

Germany's world renowned chemicals, electronics and car industries are all present in Brazil too, attracted not by financial incentives but by the sheer size of the market. BMW Brazil began manufacturing last year, and the location of a Brazilian Mercedes-Benz plant to make a new passenger car will be announced this spring.

Geographical Magazine, April 1996

The car industry is concentrated in the ABCD complex. Not surprisingly, foreign investment has been the key in this industry and others requiring large capital inputs (Figure 6.12).

In 1970, the metropolitan area reached the mark of 43.5 per cent of Brazil's industrial GNP. Since then São Paulo has had to give in to other areas.

Figure 6.13 Number employed by type of industry, Greater São Paulo

	1970	1992
Mineral extraction	2544	1383
Nonmetallic minerals	52 752	42 747
Metallurgy	122 069	154 352
Mechanics	77 857	135 462
Electrical equipment	76 424	98 125
Transport equipment	101 009	115 301
Wood	8556	7917
Furniture	30 286	24 638
Paper and cardboard	26 675	35 211
Rubber	18 865	35 504
Leather and furs	2314	2173
Chemicals	33 139	70 004
Pharmaceutical products	17 572	22 675
Scents, soaps and candles	6876	19 075
Plastics	26 932	54 595
Textiles	114 299	54 719
Clothing, footwear and cloth goods	54 000	95 984
Food products	46 829	96 150
Beverages	6453	8362
Tobacco	2903	1843
Printing and publishing	34 795	47 088
Other	33 994	44 096
Total: Greater São Paulo	**914 907**	**1 167 404**
Total: State of São Paulo	**1 295 810**	**2 208 680**
% of State total	**70.61**	**53.27**

Consequently there was a remarkable expansion to the interior of the state, especially along the Paraiba Valley and in Campinas and also to other states in the region, particularly Minas Gerais. However, little was lost overall as the movement out of traditional industries was counterbalanced by the arrival of high technology industries. Figure 6.13 shows how employment by industrial sector changed from 1970 to 1992.

More than 53 per cent of all the industries that invest in research and development are located in São Paulo. The proximity of São Paulo's three public universities, the University of São Paulo, the University of Campinas, and the University of the State of São Paulo, and the principal industrial research centres has facilitated a steady interchange of ideas and information.

Cubatão: Petrochemical Complex and Environmental Problems

Cubatão is located on an inlet close to the foot of the Serra do Mar, on the SP-150 highway between São Paulo and Santos. It has the largest complex of petrochemical plants in Brazil. In the 1980s Cubatão gained the unenviable reputation as one of the most polluted environments on earth due to a combination of negligible pollution control and local site conditions which did not favour atmospheric dispersal. Its 22 factories were belching out nearly 1000 tonnes of pollutants a day: sulphur dioxide, nitrogen dioxide, carbon monoxide, ammonia, nickel, mercury, lead and copper. The situation was so bad that the sun was frequently obliterated by the pall of stinking pollution hanging over the town. In 1984 it was estimated that each one of Cubatão's residents received a daily quota of 10 kg of pollutants. Since then the state government has compelled the polluters to treat 90 per cent of the pollution cocktail before it gets into the atmosphere. Helicopters have spread a specially-developed gelatine over the denuded hills in the surrounding area, to fix and nourish tree seeds. The UN Environment Programme presents Cubatão as an example to industrial polluters.

Cubatão has a history of less than half a century. In 1954, following Congress passing Law 2004 to establish Petrobras, the state-owned petroleum company, the company founded the town by building its Presidente Bernades oil refinery there. Other companies followed, producing steel, fertilisers, herbicides, formaldehyde, methanol, and resins.

Looking back at the Serra do Mar from Cubatão the meandering courses of the two highway routes from São Paulo, Dos Imigrantes and Ancheta, are a striking contrast to the relatively straight pipelines negotiating the scarp. The latter take oil up to São Paulo and bring water down from the plateau.

Santos: Brazil's Major Port

Founded in 1543, Santos became a major port when the railway to São Paulo was completed in 1867. With a population of 429 000 (1991) Santos now boasts one of the largest port complexes in the world. Its most important function is as the out-port for São Paulo but it is also strategically important for the whole country. Located approximately 75 km from São Paulo the port employs 40 000 people and harbours over 300 ships each month. The port is 14 m deep with 12 km of piers and 51 docks. There are 520 000 m^2 of warehouse space and 570 000 m^2 of yard space. More than 35 million tonnes of cargo such as sugar, coffee, beef and soya passed through its docks in 1995. A concern for the region's exporters is the relatively high handling charges of the port which, for example, are more than double those set by the ports of Salvador and Fortaleza. Santos's high charges are generally put down to poor efficiency. However a programme of modernisation and reorganisation is now beginning to have some impact. The port should be fully privatised by 1998. Codesp, the state port operator, predicts that lower charges and improved efficiency will double the cargo handled by the year 2000.

Manufacturing industry is very limited in Santos and outside of the docks tourism is its major function, accounting for about two-thirds of the local economy. A considerable number of São Paulo's affluent class own holiday apartments either in Santos or in nearby Guaruja. The population density in Santos doubles during the summer months (December to March) and can triple or quadruple over summer weekends.

THE ROAD TO RIO

Located just over 80 km east north east of São Paulo, on the BR 116 to Rio de Janeiro, is São Jose dos Campos, the most advanced high technology centre in South America (Figure 6.14). This is Brazil's centre for weapons, aircraft and aerospace. Its strategic location allows it to draw on a highly skilled workforce from a wide geographical area and maintain close contact with the nation's other advanced research institutions. Embraer, the Brazilian aeronautics company has produced more than 4000 airplanes in São Jose dos Campos in more than two decades of manufacturing.

Figure 6.14 The technology centres of São José dos Campos

CTA THE AEROSPACE TECHNICAL CENTER

Eight-three kilometers from São Paulo in São José dos Campos, the Aerospace Technical Center (Centro Técnico Aeroespacial – CTA) is formed by a group of five institutes. These institutes conduct operations involving teaching, research, and development in various areas of interest in aviation and outer space studies.

Connected to the Ministry of Aviation the Center was founded in 1971 and today has nearly 7,000 employees, 80% civilian and 20% military.

The major portion of the research and development done by the center for the Brazilian space program is performed in the Space Activities Institute. The program is nearly ready to launch Brazilian made satellites in Brazilian made vehicles.

Another important unit is the Institute of Advanced Studies where both pure and applied researches are performed in advanced scientific areas such as data processing, lasers, radiation-material interaction, and nuclear energy.

The Research and Development Institute works on the creation of aviation processes and products, electronic equipment, special materials and mechanical systems. The Aerospace Technical Center also furnishes services and works on information transfer with industries that can benefit from the new products and technologies developed. This interaction is accomplished through the Institute of Industrial Coordination and Encouragement. Another unit, the high education center of the Ministry of Aviation, functions with the CTA. The Aeronautic Technologic Institute is the principal Latin American Centre for training in the various modalities of engineering related to aerospace development.

Sao Paulo City Life 1993

Figure 6.15 High technology industry in São José dos Campos

One hundred and seventy five km further along the BR116 towards Rio is a total contrast – Volta Redonda, the heartland of Brazil's steel industry. Situated on the banks of the Paraiba river, the city is dominated by steel mills. To all intents and purposes Volta Redonda has been a company settlement since 1941 and its urban structure represents the priorities of the company with slums for the workers on one side of the river and carefully planned neighbourhoods for management on the other. The heavy nature of industry in the city has taken a considerable toll in terms of pollution.

No more than 40 km west of Volta Redonda is the industrial town of Resende where VW opened a truck and bus plant in 1996, bringing the state $250 million in direct investment and 1800 jobs. However, to get the factory the state had to offer an attractive package including tax breaks, free land, and VW's own facilities at Sepetiba seaport and Rio's international airport. Peugeot is currently considering Resende at its location for a $1 billion investment.

The alternative route between São Paulo and Rio is the coast road. About two-thirds of the way along this road towards Rio is Angra dos Reis, the location of Brazil's first nuclear power station. The town is also important for shipbuilding.

RIO DE JANEIRO

Rio de Janeiro's main industries are textiles, foodstuffs, cars, household appliances, cigarettes, chemicals, oil refining, leather goods and metal products. It is the country's major import port. The northernmost part of Rio, the 'Zona Norte' contains the city's industrial areas and the major working-class bairros. Industry has been pushed gradually to the periphery of the city as inner areas are increasingly devoted to services or housing.

State and federal government funding is being used to upgrade the port of Sepetiba, on Sepetiba Bay, south of Rio de Janeiro. The project emphasises the integration of industrial and port facilities with the development of inter-modal transportation. Apart from adding considerably to the state's infrastructure, the development of Sepetiba will reduce various pressures on Rio de Janeiro.

Other Industrial Centres in São Paulo State

Most settlements of any significant size in the state boast some form of manufacturing. Of particular note are:

- Campinas, 100 km north west of São Paulo which is an important agricultural processing and high technology centre;
- Baura, further north west still, a major agricultural processing centre serving the orange, pineapple and sugar producers of the area;
- Franca, in the north of the state, is the centre of the shoe industry in Brazil.

BELO HORIZONTE

Belo Horizonte, the first of Brazil's planned city's was laid out in the 1890s. As late as 1945 the population had not risen above 100 000. Today it is more than thirty times the size. In recent decades Belo Horizonte has been one of the main destinations of rural migrants, seeking work in its rapidly expanding industrial sector.The city is second only to São Paulo as an industrial centre. It lies at the centre of the rich mining and agricultural hinterland, the so-called 'Mineral Triangle', that has made the state of Minas Gerais one of the economic powerhouses of Brazil. The area contains the richest mineral deposits in the country. Today iron, bauxite and manganese have superseded the precious metals of colonial times. Belo Horizonte's chief manufacturing industries are steel, steel products, cars and textiles. The minerals of the surrounding area are also processed in the city.

Manufacturing industry is also important in a number of towns within the hinterland of Belo Horizonte, including Contagem and Itauna. The region's manufacturing companies import and export through the ports of Rio de Janeiro and Vitória. More iron ore is exported through Vitória than any other port in the world. This port city is dominated by docks, rail yards and smelters.

Figure 6.16
Belo Horizonte

Traditional Industries Filter-down to the Northeast

The shoe industry has been traditionally based in the Southeast, especially in Franca in the north of São Paulo state. But now shoe companies are beginning to shift production to the Northeast where wages are lower and government grants are available. The country's biggest shoemaker, Grendene, has already spent $100 million on four factories in Ceará and is now planning a fifth. The company claim that wage costs in the Northeast are only 36 per cent of what they are in the Southeast. Other industries such as textiles and furniture manufacture are also moving north east.

Most industries following this path are domestically oriented; the Northeast's comparatively large but poor population has been one of the main beneficiaries of the fall in inflation. The poor have been able to afford more and this has led to an increase in consumption. Such a movement of industry from one region to another seeking lower costs is known as the filter-down concept of industrial location.

? ? ? ? ? ? ? QUESTIONS ? ? ? ? ? ? ?

1. Why is São Paulo the most important industrial city in Brazil?

2. (a) To what extent did the industrial structure of Greater São Paulo change between 1970 and 1992?

(b) Suggest reasons for the changes you have identified.

3. Identify the locational factors that led to the development of industry in Cubatão.

4. Examine the reasons for industrial concentration

(a) along the BR 116; and

(b) in and around Belo Horizonte.

5. Explain how the filter-down concept of industrial location is currently operating in Brazil.

MOVING UP THE GLOBAL LEAGUE

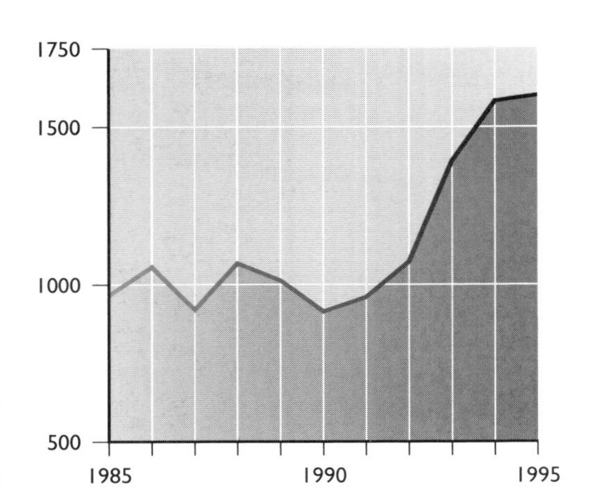

Figure 6.17 Production of cars in Brazil, in '1000s

In 1995 Brazil produced 1.6 million motor vehicles (Figure 6.17) making it the seventh largest car manufacturer in the world. Total vehicle sales in the country accounted for 65 per cent of the South American market. By the turn of the century Brazil is expected to replace Italy as the fifth biggest producer with a capacity of 2.6 million cars a year. Strong domestic demand, the advantages of Mercosul and export markets are all contributing to the industry's rapid growth. The industry, with the inclusion of parts, is expected to invest $20 billion in Brazil between 1995 and 2000.

Figure 6.18 Estimated investment 1995–2000, in the assembly of cars, trucks and jeeps

Southeast		
Fiat	Minas Gerais	3.0
Ford	São Paulo	2.5
General Motors	São Paulo	2.4
Volkswagen	São Paulo and Rio de Janeiro	2.3
Mercedes	Minas Gerais and São Paulo	0.8
Toyota	São Paulo	0.6
Honda	São Paulo	0.3
Scania	São Paulo	0.16
BMW	Undecided	0.15
South		
General Motors	Rio Grande do Sul and Santa Catarina	1.1
Renault	Paraná	1.0
Audi	Paraná	0.5
Chrysler	Paraná	0.3
Volvo	Paraná	0.15
Northeast		
Asia	Bahia	0.72
Inpavel	Pernambuco	0.3
Hyundai	Bahia	0.29
General Motors	Undecided	0.16
Subaru	Ceará	0.15
Skoda	Bahia	0.1
Troller	Ceará	0.016
North		
Zam	Acre	0.014
Nanjing	Tocantins	0.009
Centre-West		
Mitsubishi	Goias	0.035

Figure 6.18 shows the estimated investment by car manufacturers by region for the same period.

VW is the biggest car manufacturer in Brazil producing 35 per cent of total output, with Fiat coming in second with 27 per cent. The Brazilian automotive parts industry is the largest and most advanced in the developing world.

BRAZIL'S MAJOR MANUFACTURING INDUSTRY

The car industry has been the most important single element of the Brazilian economy since its introduction in the 1950s. Much of the investment in Brazil announced by multinational companies goes to the automotive sector. With a ratio of one car for every 11.3 people (Figure 6.19) the growth potential is substantial and much higher than in mature markets like Europe and the USA. Also, Brazil's current vehicle fleet, with an average age of fifteen years for buses and eleven years for cars promises a booming replacement market. The sales volume that can be achieved in Brazil is only possible in countries of continental proportions and a rapidly growing middle class – which rules out China, India and Eastern Europe, at least in the near future. Annual vehicle sales in Brazil almost doubled between 1992 and 1996 making the country one of the world's fastest-growing domestic car markets.

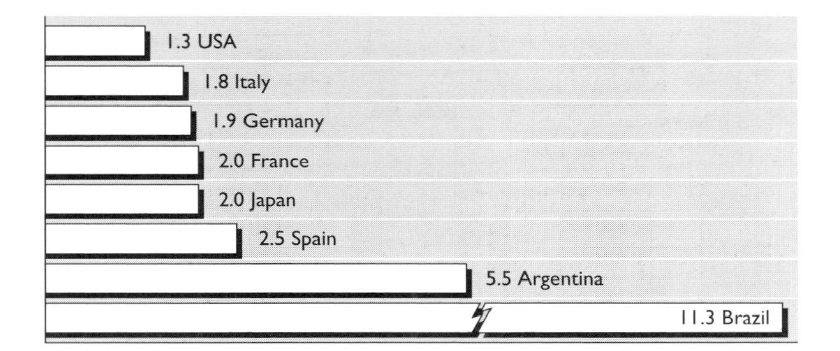

Figure 6.19 Ratio of people per car in various countries

In 1992 the government helped to kick-start the industry's growth by cutting taxes on small engined cars, which now account for 57 per cent of the Brazilian market. Since then growth has been driven by a combination of increasing competition and low (for Brazil) inflation. The latter has provided the country's lower-middle class with greater purchasing power and access to consumer credit.

Figure 6.20 General Motors car plant near São Paulo

Figure 6.21
High tech production inside a car plant

The removal of barriers to the importation of cars and parts has forced Brazil's established car manufacturers – Volkswagen, Fiat, General Motors and Ford – to modernise their outdated plants. In the past these companies had made substantial profits by producing small volumes of old-fashioned cars and selling them at high prices. The objective now is to produce the same models and engines as in their European plants and to achieve similar levels of productivity.

The productivity issue is being tackled partly by introducing 'outsourcing'. At Volkswagen's new lorry factory and new engine plant, much of the work will be done by people who work for suppliers. These workers, paid less than Volkswagen's own, will pre-assemble modular components inside the plants. The idea is that Volkswagen will be responsible for the engineering but the production will be done by suppliers. Other companies have begun to operate in a similar way. For example, Fiat has several suppliers operating inside its plant at Betim in Minas Gerais. Investment has also centred on the installation of robots. In 1990, 117 000 workers produced 900 000 cars; in 1995 104 000 workers produced 1.6 million cars.

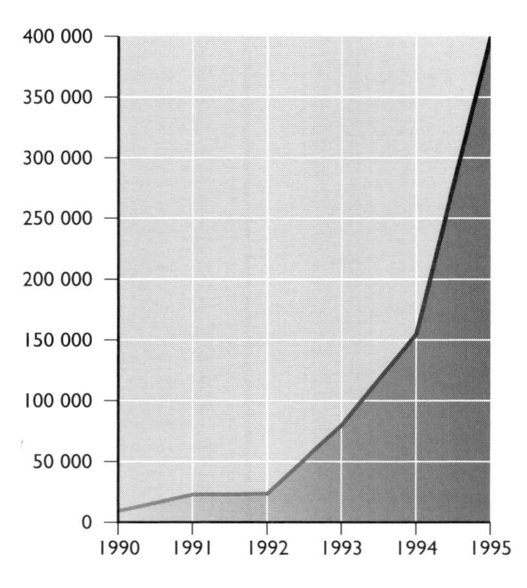

Figure 6.22
Vehicle importation, 1990–5

A major reason behind increasing capital intensity is that payroll taxes mean that a Brazilian car worker costs about $11 an hour (mid-1996), not far behind the British car industry.

In 1995, concerned by rapidly rising vehicle imports (Figure 6.22) the government once more raised the import tax to 70 per cent to stem the flow. Quotas were established but were abolished in October 1995 under pressure from the International Trade Organisation. However the Brazilian Association of Automotive Importers has agreed on a voluntary 30 per cent limit on the growth of imports per year.

INITIAL LOCATION IN THE SOUTHEAST

The first Brazilian car was a Model 'T' Ford which was assembled from imported parts in São Paulo in 1919. However, before Juscelino Kubitschek became President in 1956 only a few cars were assembled in Brazil and there was one government-owned factory manufacturing trucks. Foreign companies were invited by the Kubitschek government to establish branch plants and then pressed to manufacture more and more of the parts for their vehicles in Brazil. Motor vehicles was seen as a key industry which would stimulate the development of other industries because of the great variety of components required for the finished product.

It is not surprising that foreign car manufacturers first located in and around São Paulo. The largest metropolitan area in South America offered the following advantages:

- a good regional raw material base;
- the largest pool of skilled labour in South America;
- proximity to Santos, now one of the largest port complexes in the world;
- a mature web of industrial linkages;
- the largest market by far in the country;
- the hub of road, rail and air transport and telecommunications;
- welcoming federal and state governments.

Once the industry had become established in São Paulo it was only a matter of time before other locations in the Southeast became attractive, namely Rio de Janeiro and Belo Horizonte. For example in 1978 Fiat opened a large plant in Belo Horizonte employing over 10 000 people and producing 130 000 cars a year. However, it must be noted that the first motor vehicle plant outside the state of São Paulo opened in 1966 at Jaboatão, near Recife. Here Willys-Overland do Brasil produced jeeps and pick-up trucks.

Figure 6.23 Increasing competition for established car industries in Brazil

Vehicle manufacturing output set to soar by end of century

The big four companies with long-established operations in Brazil – Fiat, Ford, General Motors and Volkswagen – are investing heavily in the face of increasing competition from several leading Asian, American and European companies which are investing billions to establish a strong manufacturing presence in Brazil.

Among the newcomers are:

- Mercedes-Benz, who are building a $460 million plant in Minas Gerais state, which will make 70,000 of its new "A-class" cars a year from 1999;
- Asia Motors, which is planning a $500 million factory, making pick-ups and vans;
- Honda, which is investing $600 million in Brazil, including a $100 million factory in Sao Paulo state to produce 40,000 cars a year;
- Renault, which is building a $1 billion plant in Parana state to produce 120,000 medium-sized cars a year;
- Toyota, which will make 15,000 Corolla cars a year at a new $150 million plant in Sao Paulo;
- Chrysler and the Detroit Diesel Corp, which will build a $315 million assembly plant to build 12,000 pick-up trucks a year, rising to 40,000 a year;
- Hyundai, which is seeking a Brazilian partner to build a car plant as part of its planned $1.78 billion investment programme in Brazil;
- Skoda, which plans to build 5,000 trucks a year, possibly in a joint venture with a Brazilian firm. Brazil has been short-listed by BMW and Chrysler as the site of a $350 million plant to build 40,000 engines a year.

The 'big four', though, are resisting the incursions of their new competitors, and are expanding and modernising their plants to defend their 94 per cent market share.

Fiat has poured $1 billion into its Betim plant for the new Palio model, designed for use throughout the developing world. The Palio leads Fiat's impressive performance in Brazil, where it is expecting to make pre-tax profits of $400 million this year. Output at Betim is now 2,000 units a day and should reach 2,200 next year, making it Fiat's biggest plant in the world. Iveco is also investing heavily in trucks and pick-ups.

Volkswagen, too, has expanded its operations, with a truck and bus plant which it claims is the most advanced production system in the world.

Brazil [Times supplement], 9 December 1996

Although car makers are now looking beyond the Southeast in terms of location the region is still attracting substantial inward investment. For example Mercedes-Benz is building a $460 million plant in Minas Gerais state (Figure 6.23). This is only the company's second fully fledged car plant outside Germany, the first being in Alabama, USA. High labour costs in Germany are a major reason for the company's decision to extend manufacturing overseas.

THE SPREAD TO THE SOUTH

Among the new manufacturers to locate in Brazil (Figure 6.23), the most ambitious project belongs to the French company Renault which plans to invest $1 billion in a huge plant producing the 'Megane'. The site selection process promoted what the press termed a 'fiscal war' among a number of states anxious to benefit from such substantial inward investment. For Renault, locating in Brazil is a matter of survival. Ranked eight in the world with a production of 1.8 million vehicles per year, the newly privatised company lacks space to grow in Europe where competition is intense. The selection of Curitiba as the location for the new Renault plant marked an important stage in the development of the industry – the beginning of its spread outwards from the Southeast. Major factors in Renault's choice of location were:

- the high quality of life in Curitiba;
- improvements in the port of Paranaguá (Curitiba's outport) where handling charges are considerably below those of Santos;
- the success of existing multinational companies in Curitiba such as Phillip Morris, Bosch and Volvo (tractors);
- proximity to the large markets in the Southeast;
- Paraná borders Argentina and Paraguay; both viewed as expanding markets.

Figure 6.24 New employment at what cost?

The true cost of a car

The proposed fiscal bonuses offered to lure an Asian car maker into the Brazilian Northeast, issued last Friday by President Fernando Henrique Cardoso in the form of a Provisional Measure (MP) had previously leaked into the Mercosur summit in Fortaleza. As expected, not everybody was happy, and some were a little more than unhappy.

An MP is a sort of decree, which nevertheless requires approval by Congress to remain in force, and this one gives away a substantial bit of the numerous taxes applicable to the cars eventually manufactured in the north, northeast, and west-central Brazil, up until 2010. These incentives seem powerful enough to bring in the factories — previous efforts have failed so far — and a measure of their powers is the fact that an average of 34% of the consumer-end price of a Brazilian automobile represents tax.

Brazilian taxpayers are therefore supposed to pay for a good portion of every car rolling out of there for a few years. Manufacturers who buy into the region will be exempt from Industrialized Products Tax (IPI, federal) as well as the Import Tax on machinery and equipment, tools, moulds and other components necessary to build the lines.

In other areas, the Import Tax has been reduced by 90%, while the IPI will remain unaffected. What's more, the new measure exempts financing of imported goods by means of foreign exchange operations from the Financial Operations Tax (IOF). Companies that pre-qualify between the date the MP is published and March 31, 1997 will be eligible for the benefits.

The measure has sparked a political powderkeg. Opposition stems from the fact that the incentives strongly favor the northeastern state Bahia, which competes with the other states for the installation of Asian automobile factories. Consequently, detractors suspect the measure was created specifically to gain re-election support from Senate President Antônio Carlos Magalhães.

Gazeta Mercantil, 23 December 1996

More recently the state of Rio Grande do Sul has been criticised by competing states for offering an over-generous package of incentives to General Motors. GM was given R$253 million before plant construction had even started. Opponents argue that:

- the state is failing to collect taxes from those who can afford them;
- the capital intensive nature of the modern car industry brings only limited new employment;
- Major car manufacturers have become adept at playing off one state against another, considerably reducing the benefits of location to the chosen state and the country as a whole.

NEW PLANTS PLANNED IN THE NORTHEAST

Foreign vehicle manufacturers have also targeted the Northeast in terms of car plant location. This has been due mainly to government incentives aimed at encouraging the industry to spread out beyond the Southeast and South (Figure 6.24). Asia Motors should be the first car manufacturer to benefit from the incentives included in the new Provisional Measure (MP). Planning to build either in the state of Bahia or in Ceara an investment of $500 million should realise production of 60 000 cars by 1999.

While labour availability is high in the region, levels of education and industrial skill are low compared to the Southeast and South. Thus low productivity appears to be the major obstacle for the car industry to overcome in this problem region.

STAGES IN DEVELOPMENT

Figure 6.25 can reasonably be seen as the life cycle of the Brazilian car industry. The country has recently moved into stage four as exemplified by GM's São Jose dos Campos plant near São Paulo which exports engines to the USA. Although some cars made in Brazil have reached markets in the Developed World, the total number is very small. The eventual aim shared by the government and leading manufacturers is that a considerable proportion of cars produced in Brazil will be shipped to the developed world, a purpose for which the Northeast is strategically located. When stage five is achieved Brazil will really have come into her own as a global player in the industry.

Figure 6.25 Stages in the development of the Brazilian car industry

Completed

1. Foreign multinationals assembling components mainly produced in developed countries for the Brazilian market.
2. Foreign multinationals assembling components mainly produced in Brazil for the Brazilian market.
3. Foreign multinationals assembling components mainly produced in Brazil for the South American market in general.
4. Foreign multinationals exporting parts to developed countries.

Future

5. Foreign multinationals exporting cars to developed countries.
6. Brazilian car manufacturer(s) compete for the domestic market with foreign multinationals.

CAR PARTS MANUFACTURERS

There are 548 car parts manufacturers in Brazil, the great majority of them in São Paulo state. However the decentralisation of the car assembly industry is beginning to have an impact on the location of suppliers, for example Fiat's decision to build a new factory in Minas Gerais. Fiat will use Just-In-Time manufacturing techniques and is insisting on all its suppliers being not more than 120 km away from its factory.

The car parts industry employs 214 000 people. Most of the firms are small with 72 per cent of them employing fewer than 500. Only 12 per cent have more than 1000 employees.

? ? ? ? ? ? ? QUESTIONS ? ? ? ? ? ? ?

1. Examine and comment on the four stages in the locational history of the car industry in Brazil.
2. Why is the market for cars growing more rapidly in Brazil than in other large developing countries such as India and China?
3. Comment on the logic of Figure 6.25. How many other stages, if any, could you add to the model?
4. Explain the following statement: 'The car industry has a considerable multiplier effect leading to the growth of other industries'.
5. Why is the location of car parts manufacturers becoming more firmly tied to the location of car assembly plants?

7

AGRICULTURE

An Overview

A MAJOR EXPORTER

Brazil is a major world exporter of foodstuffs. With its great size and climatic diversity it is able to produce a wide variety of agricultural products ranging from tropical products in the north to temperate foods in the south. Agriculture is an important sector of the economy and a significant element of the balance of payments (Figure 7.1). Brazil's agricultural GDP increased by an average 3.2 per cent between 1990 and 1994 compared to an average of 1.7 per cent for the rest of South America.

Figure 7.1 Brazil's agriculture helps balance the trade deficit

Farm products boost exports

Farm products account for a much smaller share of B-azilian exports today than they used to. Coffee, for instance, once provided more than 80% of the country's exports. Today it amounts to less than 5%: green coffee shipments totalled $1.7 billion in 1996, while overall exports rang up $47.4 billion.

Despite their reduced share in exports, farm products have been helping to hold the expansion of the Brazilian trade deficit, which may well jump to between $8 billion and $10 billion this year, up from 1996's $5.3 billion.

Prospects of a good harvest are a real relief for the government, as agricultural commodities currently account for 25% of overall Brazilian exports (see table below).

The government, by exempting exports of agribusiness products from Value Added Tax (ICMS), also boosted the industry.

A reflex of the ICMS exemption was the performance of soybean exports, Brazil's main farm commodity today. Projections are that revenue from soybean exports will be as high as $5.1 billion this year, a 15% increase over last year's $4.43 billion.

Coffee itself is expected to post a solid growth this year. According to projections by the sector, sales of green coffee and instant coffee should total $2.33 billion in 1997, 11.4% up from last year's $2.09 billion.

Not only do increased exports help reduce Brazil's trade gap, but they're also raising rural income. This year, some $2 billion will likely be added to farmers' income, to reach $15 billion.

ITEMS	% OF OVERALL EXPORTS	VALUE IN 1996 (IN MILLIONS OF DOLLARS)
Raw coffee grains	4.67	1718
Tobacco leaves	1.65	1028
Soybeans and soy meal	5.95	3744
Chicken	1.37	839
Crude soy oil	2.22	685
Chemical wood paste	3.11	999
Granulated sugar	2.24	934
Orange juice concentrate	2.38	1391
Instant coffee	0.97	376
Total	24.56	11 718

Gazeta Mercantil, 10 March 1997

Nevertheless, until recently Brazilian agriculture suffered because of the government's preoccupation with industry. The EIU's Country Profile for 1994–5 described the agricultural sector as a 'victim of neglect' stating that 'in many parts of the country little attention is given to fertilising, seed selection or effective crop rotation and techniques are backward'. Large scale mechanisation has until comparatively recently been confined to the southern states of São Paulo, Paraná and Rio Grande do Sul.

CROPS

Figure 7.2 shows the national planted area and production for each major crop. Grain production reached a new record in 1995 (Figure 7.3), guaranteeing domestic supply of food staples. These impressive results are largely due to increased investment in new technology. Brazil trails only the USA in the production of soya with yields above the world average but 6.5 per cent behind those achieved in the USA. The soya boom began in the 1970s. The leading states of Rio Grande do Sul, Mato Grosso and Paraná account for approximately two-thirds of production. The expanding global market for soya has made it a popular crop in agricultural frontier areas. Farmers in Amazonas will harvest their first crop of about 300 tonnes in 1997, increasing to an expected 20 000 tonnes the following year. An important stimulus is the new Madeira-Amazon waterway which has cut transport costs by up to 20 per cent.

Figure 7.2 National planted area and production by product

CROP	1996 ('000 HA)	1995/96 ('000 TONNES)
Cotton	820 552	1 023 384
Rice	3 921 256	10 038 838
Potatoes (1st crop)	109 509	1 520 020
Potatoes (2nd crop)	62 021	769 621
Cocoa	729 443	312 939
Coffee	1 982 045	2 527 749
Sugarcane	4 818 114	324 413 541
Onions	74 676	944 431
Beans (1st crop)	2 711 452	1 332 847
Beans (2nd crop)	1 944 115	24 569 111
Maize (1st crop)	11 644 066	28 366 529
Maize (2nd crop)	1 763 655	3 644 119
Soybeans	10 725 907	23 170 945
Wheat	1 831 521	3 276 945

Figure 7.3 Grain production, 1990–5

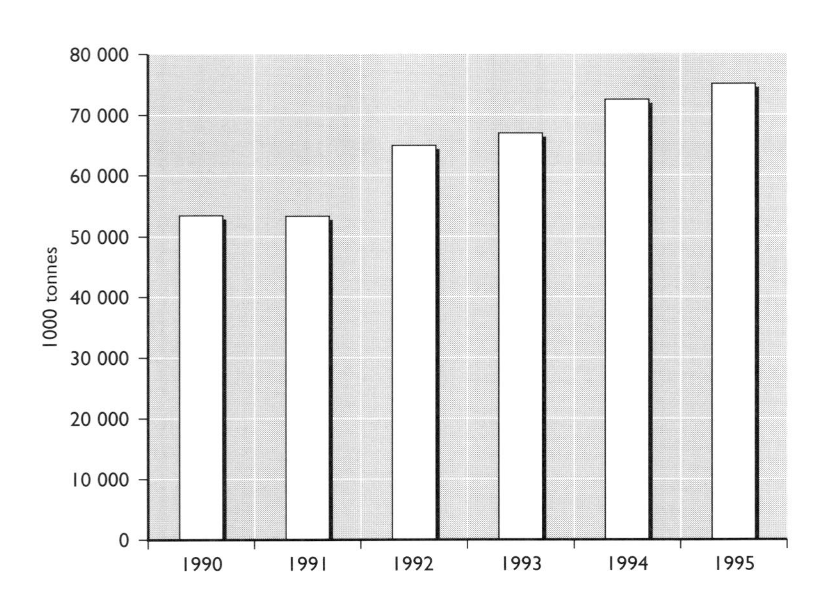

However, yields are not as high for all grains. Brazil's corn yield averages 3 tonnes per hectare compared to 10 tonnes per hectare in the USA. While corn production methods are very advanced in the prime agricultural belts such as the Alta Mogiana region in the interior of São Paulo state and the Ponta Grossa region in Paraná, in many other areas traditional methods are still practised. Yet even in the latter areas things are beginning to change.

Improved farm income from record harvests and new government investment in infrastructure are encouraging investment in agricultural technology. The stable economic situation has stimulated domestic demand such that the per capita consumption of grain in Brazil increased from 359 kilos in 1994 to 377 kilos in 1996.

Brazil is the world's largest sugar producer (Figure 7.4) and is easily the most important exporter. Almost 70 per cent of the crop is processed into alcohol for fuel with the rest destined for sugar production. Over 60 per cent of the country's sugar emanates from the state of São Paulo.

Figure 7.4 Sugarcane production, area and yield; leading countries, 1995

COUNTRY	PRODUCTION 1000 TONNES	AREA 1000 HECTARES	YIELD IN KILOS PER HECTARE	% OF WORLD PRODUCTION
Brazil	295 161	4405	67.0	26.1
India	259 490	3750	69.2	23.0
China	61 000	1000	61.0	5.4
Thailand	50 597	923	54.8	4.5

Brazil's best known export product, coffee, accounted for no less than 70 per cent of total exports in the 1930s. Its contribution has declined since then, particularly since the 1950s, with the increasing diversification of the Brazilian economy. Today coffee is responsible for less than 5 per cent of total exports and now ranks only seventh among farm products in terms of value. Nevertheless Brazil remains the world's number one producer (Figure 7.5) in spite of adverse climatic conditions reducing 1995 production by 30 per cent compared to the previous year. Together with the other members of the Association of Coffee Producing Countries, set up in 1994, Brazil agreed to several measures designed to support international prices. Over half of Brazil's coffee comes from the state of Minas Gerais.

Figure 7.5 Coffee production, area and yield; leading countries 1995

COUNTRY	PRODUCTION IN 1000 TONNES	AREA IN 1000 HECTARES	YIELD IN KILOS PER HECTARE	% OF WORLD PRODUCTION
Brazil	1819	1816	1001	28.0
Colombia	810	950	853	12.5
Mexico	408	772	528	6.3
Indonesia	346	773	448	5.3

Brazil also leads the world in the production of oranges (Figure 7.6), the country's ninth ranking farm product by value. Three-quarters of the crop comes from the state of São Paulo. Apart from sustaining traditional agricultural practices, the government has been keen to encourage new sectors of the industry. Such is the case with the production of orange juice which did not exist in Brazil 30 years ago. FMC, the world's leading manufacturer of complete turn-key citrus processing systems is located in Araraquara, in São Paulo state. The main problem for orange juice exports is the high US import tariff ($467 per tonne of juice).

Brazil ranks as the largest cocoa producer in South America and third in the world. Over 80 per cent of the crop comes from plantations located in Bahia with over half of production destined for export markets.

Figure 7.6 Orange production, area and yield; leading countries 1995

COUNTRY	PRODUCTION IN 1000 TONNES	AREA IN 1000 HECTARES	YIELD IN KILOS PER HECTARE	% OF WORLD PRODUCTION
Brazil	19 693	828	23.8	34.1
United States	10 538	288	36.6	18.2
Mexico	3549	221	16.1	6.1
Spain	2439	137	17.8	4.2

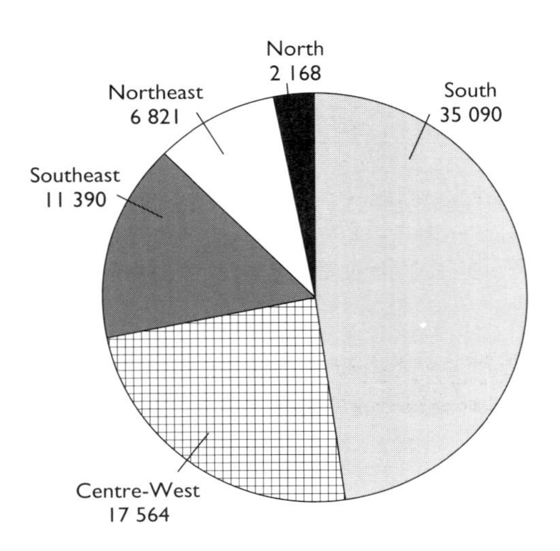

Figure 7.7 The distribution of total production of cereals, vegetables and oilseeds, million tonnes

Beans, together with rice and manioc, are the three most important items in the staple diet of most Brazilians. Thus although Brazil is the second largest producer in the world it does not export beans. As with rice (ninth in the world), all that is produced is consumed by the domestic market. Cultivation of beans is widespread across the country and comes from smallholdings. Brazil is eighth on the world list of cotton producers.

Figure 7.7 shows the regional distribution of cereal, vegetable and oilseed production by region. The dominance of the South compared to the limited role of the North and Northeast combined is the most noticeable fact.

LIVESTOCK

Beef is the most important farm product by value in Brazil. Production of 4.6 million tonnes in 1995 was second only to the USA in the world ranking. However yield is low and trails the global average. Only about 10 per cent of production is destined for the export market which is still very protectionist in nature. Although Brazil ranks seventh in the world for milk , productivity is low compared to Europe and North America. Over 30 per cent of production is concentrated in the state of Minas Gerais.

The livestock product registering the most spectacular gains in recent decades is chicken. With a 1995 production of 3.8 million tonnes (15 per cent exported) Brazil trailed only the USA and China. Once a luxury, chicken has now become a regular part of most Brazilians' diet. Chicken consumption increased from 2 kg per person in the 1970s to 22 kg per person in 1996. Total consumption of animal protein per person rose from 54 to 64 kg from 1994 to 1996.

RECORD SALES AND INVESTMENT

Sales of fertiliser, herbicides, pesticides, seeds and animal feed totalled more than $10 billion in 1996. In addition farmers benefited from the government's decision to abolish value added tax on agricultural exports (Figure 7.1), which corresponds to an additional income of $ 2 billion a year.

Rising levels of production have had much to do with changing attitudes in the farming community. Prior to 1994 the agricultural sector's inefficiencies were largely masked by high inflation. Many farmers made more money through speculation than actual farming. In the new era of economic stability farmers are focusing more firmly on productivity.

Brazil's major food companies and farm cooperatives are investing heavily. In the 1990s only the chemical industry has benefited from a higher level of investment than the food sector. Farm sector income is predicted to grow from $63 billion in 1996 to $75 billion in 2000.

NEW AGRICULTURAL POLICY

A new agricultural policy was introduced with effect from the 1996–7 harvest to guarantee a higher income for farmers and an increased supply of food to the public. It provides more resources for rural credit, lower interest rates and a simplification of procedures for financing and marketing, putting more emphasis on the market and improvements to infrastructure.

In order to make a variety of services and adequate infrastructure available to farmers, the government, in partnership with private enterprise, is developing regional projects such as the establishment of multi-use transport corridors and the expansion of the technical assistance programme. Nineteen ninety-seven saw the opening of the first multi-use corridor linking the states of Mato Grosso and Rondônia to the port of Itacoatiara in Amazonas and from there to the Atlantic Ocean as an export corridor.

? ? ? ? ? ? ? QUESTIONS ? ? ? ? ? ? ?

1. Assess the importance of agriculture to the Brazilian economy.
2. Comment on the relationship shown in Figure 7.2 between planted area and production.
3. Give reasons for the significant increase in the production of grain in recent years.
4. To what extent is Brazil an important contributor to the global production of farm products?
5. Suggest reasons for the regional distribution shown in Figure 7.7.
6. How will the development of multi-use transport corridors benefit farmers, particularly in frontier areas?

Land Reform

A PROBLEM SINCE COLONIAL TIMES

The distribution of land in terms of ownership has been a divisive issue since the colonial era. Then the monarchy rewarded those in special favour with huge tracts of land, leaving a legacy of highly concentrated ownership. For example 44 per cent of all arable land in Brazil is owned by just 1 per cent of the nation's farmers while 15 million peasants own little or no land. Many of these peasants are impoverished, roving refugees who have lost their jobs as agricultural labourers due to the spread of mechanisation.

The system of land ownership in Brazil and many other developing countries is one of the major causes of poverty. It is a feudal style system, similar to the pattern which existed in Britain in the Middle Ages. The answer to this problem is land reform. This would involve breaking up big estates and redistributing the land to peasants. According to the Catholic Church's land commission 49 000 landowners each hold an average of 3347 hectares.

Although successive governments have vowed to tackle the problem, progress has been very limited due to the economic and political power of the big 'fazenda' or farm-owners. The latter have not been slow to use aggressive tactics, legal or otherwise, to evict squatters and delay expropriation. Their actions have included bribing and intimidating members of the judiciary and land-reform officials in order to keep their large estates intact.

VIOLENT CLASHES

In the mid-1990s land reform clearly emerged as Brazil's leading social problem, highlighted by a number of widely publicised squatter invasions. Such land occupations have occurred in both remote regions and established, prosperous farmlands in the South and Southeast. Even the capital city has not escaped. In May 1996, 300 families occupied a plot of public land outside Brasília.

Figure 7.8 Land reform demonstration

The most organised occupations are led by sophisticated groups of peasants and sympathisers (left-wing politicians, trade unionists and Catholic clergy), known as the MST, Movimento Sem Terra. While most occupations are peaceful, more than 200 peasants have been killed by police or by hired guns in land conflicts around the country. Very few prosecutions have resulted, to the great indignation of the landless. However, two of the worst incidents have occurred very recently. In August 1995, nine people were killed when 180 military police stormed an emcampment near Corumbiara, in the western Amazon state of Rondônia. Even more died on 17 April 1996 after landless peasants blocked a highway in Eldorado de Carajás (Figure 7.9).

Figure 7.9
What it costs to start land reforms

Of Land and Death

SOME 1,500 PEASANTS FROM NORTHERN Pará state in the Brazilian Amazonia wanted land, and they were hungry enough, desperate enough, to take bold action to get it. On April 17 they blocked a highway in Eldorado de Carajás to draw attention to their demand for the right to settle on idle farmland nearby. To their consternation the state government responded with busloads of heavily armed military police. After the cops fired a volley of tear gas, the peasants charged, waving machetes, hoes, scythes and a few pistols. The police opened fire with automatic weapons.

The result was the bloodiest confrontation in the 30-year history of Brazil's land-reform movement. Nineteen demonstrators died and 40 more were wounded by the police fusillade. The scene, filmed by a local television newsman and broadcast repeatedly in the following days, stunned all of Brazil. It also galvanized the government of President Fernando Henrique Cardoso to take long-overdue steps to ameliorate ancient injustices.

Saying he was ''still shocked'' by the Eldorado de Carajás incident, Cardoso last week announced that he would create a separate Ministry of Land Reform. Justice Minister Nelson Jobim followed up by pressing debate in Congress on bills that would make it more difficult for farm owners to get eviction notices against peasants squatting on their land, and he would speed up government expropriation of unused private land. Cardoso also proposed a new civilian police force to handle land conflicts and urged the Senate to speed approval of a bill, already passed by the lower house, that will give civilian courts jurisdiction over crimes committed by the military police.

Cardoso, a left-leaning sociologist who has been an advocate of social justice for decades, is under immense pressure to push through laws that will for the first time make land reform a reality. But the counterpressure is strong; agribusiness interests directly control 175 of the 513 votes in Brazil's Chamber of Deputies and can influence many more.

Time, 6 May 1996

Figure 7.10
Is land the answer?

Thanks to the machines and methods of modern farming and ranching, millions of small farmers have been tossed off their plots in the past three decades. The resultant jetsam streamed into the backlands, to the frontier lands of the Amazon or the western plains. Or into the cities, to crowd and swell the wretched slums.

Modern economics carried on what modern agriculture had begun. For decades Brazil gave protection to home-produced goods, from shoes to soya beans. Now it has opened its borders. The result? At least 500,000 countryside jobs have gone since 1990.

Most have gone for ever. And big producers say that seizing the best farm land and parcelling it out to the landless millions would make matters worse, provoking a collapse of food output, ''We are confusing a social problem with an economic one,'' says Geraldo Muller, a Sao Paulo sociologist. ''Land cannot be used as a kind of dole for the jobless.'' What Brazil really needs, say Luiz Marco Hafers, president of Rural Society, a producers' association with 12,000 members, is not land reform but farm reform. ''When farmers with 30 or 40 years experience are leaving the land because they can't make ends meet, it's fantasy to imagine that someone just arriving will be able to produce efficiently.'' Maybe. The *sem terra* would be happy to be allowed to produce at all.

The Economist, 13 April 1996

The reduction in farm employment due to increasing mechanisation has been occurring for some time but more recently the operation of international trade has made the problem worse. For decades Brazil protected home-produced goods but has recently reduced its trade barriers as part of GATT, the General Agreement on Tariffs and Trade. However, one result is that even more farm jobs have gone (Figure 7.10).

RECENT LEGISLATION

Brazil's agrarian reform initiatives settled more than 42 000 families in 1995 (Figure 7.11). Moreover, the National Program for Strengthening Family Farming gives assistance to the poorest rural families because it is not enough just to settle them on the land.

Figure 7.11 Families settled with government aid

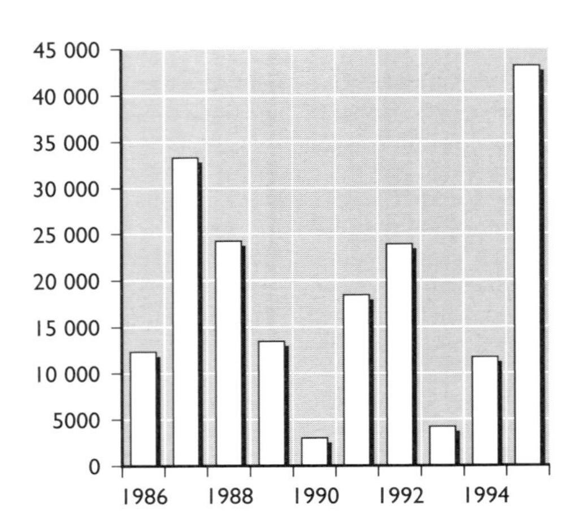

In a new move President Cardoso announced in November 1996 that a rural land tax is to be introduced to speed up land reform. The ITR tax is mainly aimed at landowners of unproductive properties who will have to pay an annual tax of up to 20 per cent on the value of their land. Land considered unproductive, and whose owners do not pay the new tax, can be confiscated by the state after five years and included in the agrarian reform programme. The MST has given a muted welcome to the government's proposal.

Figure 7.12 The fall of land prices

Land prices collapse

Land prices, which have been collapsing over the past two years in Brazil, are expected to fall even further after the approval by Congress of the new rural land tax (ITR) last week.

Forced down by as much as 50% in the Central-West and 30% in the South of the country because of the Real currency stabilization, unused rural properties have now been hit by yearly assessed taxes of up to 20% of their declared values.

Property taxes on land effectively used for agricultural purposes, have remained unchanged at between 0.03% and 0.045%.

Other contributing factors for the collapse of rural land prices are lower inflation—which diminished the value of land as a safe investment or speculative reserve—as well as globalization, which forced farmers to invest in productivity, quality and technology, rather than in large properties.

Another reason for the tumbling prices is that land offer has largely outstripped demand, as farmers are wanting to sell off considerable tracts to pay off debts and corrections of agricultural financing contracts and because of low produce prices on the market.

Large economic, industrial and financial groups, which before had large interests in ranching, are also pulling out to concentrate on their core business which is demanding more attention as a result of increased competition caused by globalization.

Gazeta Mercantil, 23 December 1996

The Minister for Agrarian Reform, Raul Jungman said the receipts from the ITR tax would increase to $1.5 billion by 1998 with all the money going directly to agrarian reform. The new tax is one of the reasons for the recent considerable fall in the price of agricultural land (Figure 7.12).

Liquidated banks have become a new source of land for agrarian reform. In September 1996 the National Land Reform Agency (Incra) announced the purchase of 24 417 hectares of land, the assets of the failed Banorte and Banco Economico. The Ministry of Agrarian Reform plans to place 1000 landless families on this newly acquired land.

THE ECONOMIC RATIONALE

The 1996 Human Development Report quotes a land survey in Northeast Brazil which showed that farms of 0–10 hectares had production of $85 per hectare, while the largest farms, those over 500 hectares, had a gross output of only $2 per hectare. The reasons identified for this relationship are:

- as farm size increases, the proportion of land in productive use declines; and
- there is an inverse relationship between farm size and the amount of labour used per unit of area.

The Report concludes that an agricultural development strategy centred on small farms rather than large simultaneously increases the social efficiency of resource use in agriculture and improves social equity through employment creation and the more equal income distribution that small farms generate.

? ? ? ? ? ? ? QUESTIONS ? ? ? ? ? ? ?

1. Briefly explain the historical background to the issue of land reform.
2. Suggest why it is justifiable to refer to land reform as the most pressing social issue facing Brazil today.
3. Why has it proved so difficult to legislate and enforce such legislation?
4. To what extent might there be a conflict between the social justice objective of land reform and the economic reality of the issue?
5. (a) Why should the new tax on unproductive land speed up the process of land reform?

   (b) Discuss the reasons for the recent fall in the price of agricultural land in Brazil.
6. What is the economic argument for land reform?

Paraná: An Advanced Agricultural State

No other sector of the Brazilian economy exhibits the gulf between rich and poor more than agriculture. While millions of peasant farmers can manage little more than to scratch a living from the land, paricularly in the Northeast and North, the most prosperous agricultural areas operate to a degree of sophistication similar to that in the most advanced farming nations. Brazil's most advanced farms are concentrated in the South and Southeast where the term agribusiness has become increasingly appropriate to describe their internal organisation, productivity and links to other food sectors.

A MAJOR FOOD PRODUCER

Paraná is the fifth most important state in Brazil in terms of total economic output. Its physical and climatic variations allow the pursuit of different types of agricultural activity and its level of economic development means that advanced agricultural techniques are used, resulting in the highest productivity rates in Brazil (Figure 7.14). In 1993 it was the largest producer of cotton, corn and wheat, and runner-up in soya beans and common beans production. Corn and soya occupy by far the largest areas of farmland in the state (Figure 7.15). Paraná ranks first in the country in terms of overall grain production. Paraná also records the largest pig population in the country (11 per cent of the country's total), as well as an appreciable number of cattle (6 per cent of the total national herd).

Figure 7.13 A team of modern combine harvesters in action

Figure 7.14 Characteristics of farming in Paraná (compared to Brazil as a whole)

1. Large average family farm size
2. High degree of mechanisation and automation
3. Large input of fertiliser, herbicide and pesticide
4. High productivity per agricultural worker
5. Significant proportion of farmers with agricultural qualifications
6. Overwhelming dominance of cash crops
7. Large export ratio
8. Efficient distribution network
9. Well organised system of marketing
10. Strong linkage between the primary, secondary and tertiary sectors of the food industry

The state also ranks as the third biggest poultry producer (15 per cent of national production). Milk production in Paraná is about 10 per cent of the total national production.

Other important farm products in Paraná are potatoes, sugar cane, cassava and rice. In recent years, programmes have been set up to develop fruit growing in various parts of the state. In the northern part of Paraná, the planting of citrus orchards has led to industrial production of orange juice. Apples are grown in various regions giving an annual production of around 30 000 tonnes. Tropical fruits are grown in the coastal region.

Figure 7.16 shows the value of the different agricultural sectors in recent years. However, following the usual process of economic development it is not surprising that the contribution of agriculture to the economy of the state has been falling in recent decades (Figure 7.17).

Figure 7.15 Agricultural production in Paraná State, 1992–4

PRODUCTS	1992-3		1993-4	
	HARVESTED AREA (1000 HA)	PRODUCE (1000 TONNES)	HARVESTED AREA (1000 HA)	PRODUCE (1000 TONNES)
Sugar cane	196	14 000	216	15 500/16 500
Corn	2703	8158	2830	7800/8600
Soya	2076	4817	2180	4870/5280
Wheat	696	1023	700	1280/1400
Manioc	137	3014	168	3500/3800
Coffee	230	100	200	80/100
Cotton	345	448	239	415/450
Potatoes	41	625	27	408/447
Beans	546	444	577	478/483
Rice	128	233	120	226/247

Figure 7.16 Gross agribusiness production in Paraná, 1993 and 1994 by value

GROUPS	1993 (IN US$ 1000)	%	1994 (IN US$ 1000)	%
Grains & cotton	2 165 436	51.5	2 293 327	51.9
Soya	871 599	20.7	907 028	20.5
Corn	824 196	19.6	752 045	17.0
Coffee	94 939	2.3	98 530	2.2
Winter grains	129 721	3.1	160 616	3.6
Wheat	118 238	2.8	149 001	3.4
Vegetables	144 765	3.4	216 637	4.9
Others	342 936	8.1	404 140	9.2
Livestock	1 424 852	33.9	1 345 497	30.4
Total	**4 207 710**	**100.0**	**4 420 217**	**100.0**

Figure 7.17 Internal gross product, Paraná 1970–89

YEAR	FARMING/LIVESTOCK	INDUSTRY	SERVICES
1970	25.6	23.6	50.8
1980	19.4	28.8	51.8
1989	14.0	26.3	59.7

Almost 50 per cent of state income derives from activities directly or indirectly connected with agribusiness. The presence of a significant number of manufacturers of farm machinery, fertilisers, pesticides, and herbicides is a clear indication of the standing of the farm sector in the state. In terms of agricultural machinery the major producer in the state is New Holland. In 1993 Paraná accounted for 17 per cent of all tractor and harvester sales in Brazil.

The agribusiness complex accounts for 60 per cent of Paraná's exports. Soya products alone make up 39.45 per cent of exports (28.04 per cent soya bran, 8.02 per cent soya grains, and 3.32 per cent soya oil). Poultry and coffee exports account for 6 per cent and 4.7 per cent respectively. However, in contrast to the general trend, cotton exports decreased from $82 million in 1991 to $9 million in 1993 as production became less profitable due to the abolition of the cotton import tax.

HIGHER VALUE PRODUCTS

The last decade has witnessed the steady move to higher value agricultural products such as the processing of instant coffee, refined soya oil, corn by-products, animal feedstuffs, milk-based desserts and preserves. The coffee industry in the state is now targeting the fast-growing gourmet market which is presently dominated by other coffee growing countries such as Colombia and Costa Rica.

One of the major players behind the push to higher value production has been the large cooperatives such as the Organicao dos Cooperativos de Paraná. Such is the size and strength of these organisations that there are signs that they might start to imitate other cooperatives around the world which have begun to open themselves to private capital.

Figure 7.18 Large scale production of soya

COTTON

Cotton growers have had a difficult time in recent years. In 1992, Paraná had 705 000 hectares under cotton. However this fell to 63 000 hectares in 1996 as Brazil reached the status of one of the world's largest cotton importers from its previous status as the world's leading exporter. The change was mainly due to a shift in government policy which opened the door to unrestricted imports which enjoyed financing terms ranging from six months to a year at interest rates averaging 2.5 per cent a year. Acutely aware of the perilous situation of domestic producers, the government introduced Provisional Measure 1569 in early 1997 which banned short-term credit for imports. As a result the area under cotton in the same year should rise to 100 000 hectares and increase further in the following years. The cotton industry is resuming production at the same time that major technological advances are being achieved. It is not surprising that those growers who managed to stay afloat when cotton production contracted were those who had invested significantly in mechanisation.

THE LIME APPLICATION PROGRAMME

One of the problems hindering productivity in some parts of the state is the acidity of the soil. Paraná increased lime application from 3 million tonnes in 1994 to 6 million tonnes in 1996, targeting small and medium size farms in particular by subsidising the product to farmers. The construction of 38 road terminals to shorten the distance between lime production sites and farming areas has been an important element of the initiative.

PARANAGUÁ

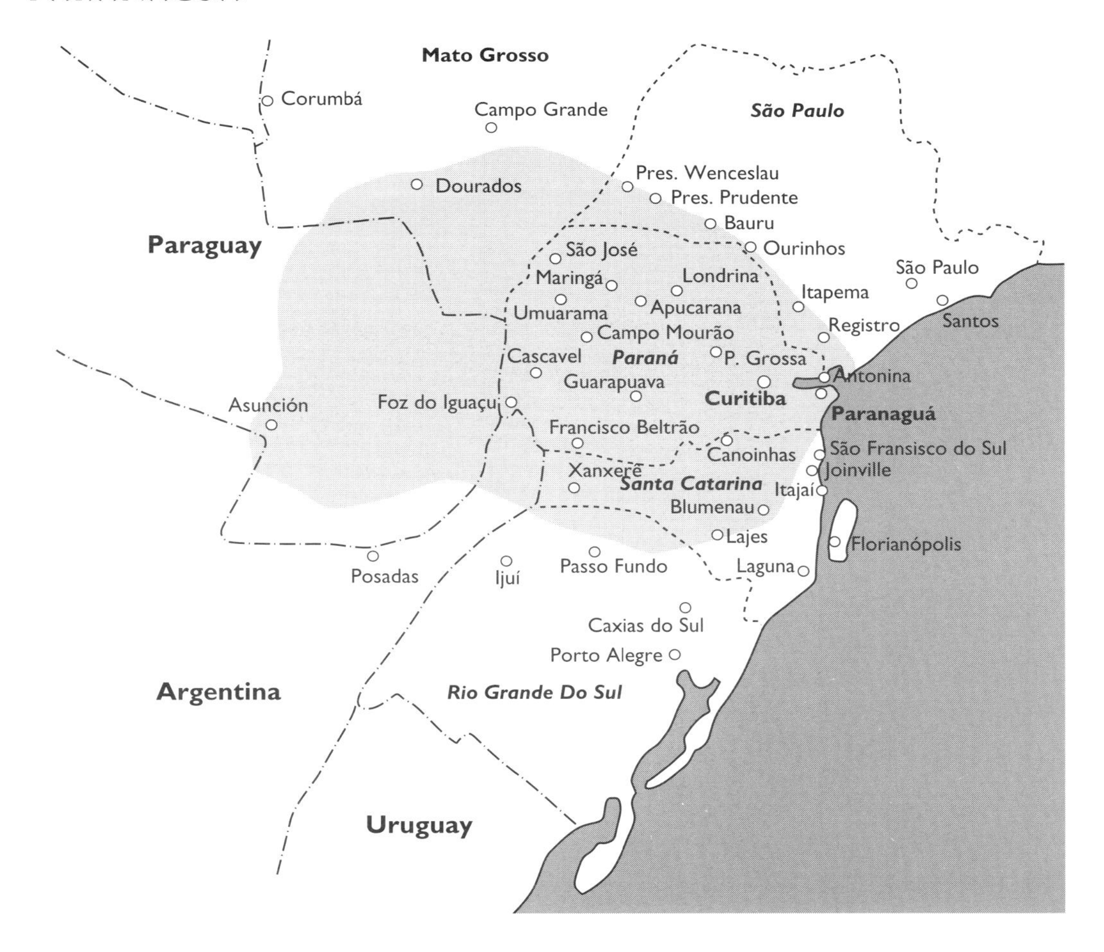

Figure 7.19 Paranaguá harbour, position, characteristics and equipment and region of influence

The functioning of an efficient agribusiness sector relies heavily on good systems of infrastructure. The port of Paranaguá (Figure 7.19), 90 km along a four-lane highway from the state capital of Curitiba, is a vital part of the state's transportation system. It is one of the best equipped and most efficient harbours in Brazil. Its sphere of influence extends beyond Paraná itself to include the state of Mato Grosso do Sul and Paraguay. The port receives cargoes from trucks, trains and ducts. In 1993 the port processed 15 million tonnes of cargo. Paranaguá is the largest grain exporting seaport in South America and Brazil's third largest in overall cargo handling. It is able to receive vessels up to 70 000 tonnes. Facilities for grain and bran warehousing provide storage for almost 100 000 tonnes. Reception, weighing, storage, forwarding and shipment of grains and bran are carried out mechanically and organised from a central control complex. A container terminal, which will add significantly to the port's facilities, is under construction.

OTHER ELEMENTS OF INFRASTRUCTURE

The state has 15 000 km of paved roads (and 245 000 km of unpaved roads), giving it a better road transport system than most other parts of Brazil. The railway network (3300 km) is not extensive but there are a number of important links, for example, between Curitiba and Paranaguá. Paraná is a large producer of electricity, led by the huge Itaipú power plant. Over three-quarters of the electricity produced is sold outside the state. Curitiba claims the most advanced telecommunications system in Brazil.

? ? ? ? ? ? ? QUESTIONS ? ? ? ? ? ? ?

1. Select six production figures to summarise the importance of agriculture in Paraná.
2. Explain the shift in production to higher value agricultural products.
3. Identify the various elements of 'agribusiness'.
4. Why is infrastructure so important to the efficient functioning of agribusiness in the state?

The broad picture

AN UNDERDEVELOPED INDUSTRY

Although Brazil now counts on services for half its GDP, tourism is one part of this sector which is relatively undeveloped, thus offering considerable scope for expansion in the future. Nevertheless, tourism does currently represent as much as 7.8 per cent of GDP, and employs, directly and indirectly, about six million people.

Figure 8.1 Balance on Brazil's international travel account 1992–4, US$ million

	1992	1993	1994
Receipts	999	1043	944
Expenditure	1318	1842	2156
Balance	−319	−799	−1212

The late 1980s was a gloomy period for international tourism in Brazil as the number of foreign visitors fell from 1.7 million in 1988 to little over 1 million in 1990. However, since then the situation has steadily improved, with 2 million foreigners arriving in 1994 and spending a total of $1.8 billion. Not surprisingly, Brazil trails the major European countries such as Spain (43 million visitors, spending $21 billion), but it is also some way behind Latin American rivals such as Mexico (16.4 million visitors, spending $16.8 billion) and Argentina (4 million visitors and $4 billion). Although rising prosperity is encouraging more Brazilians to take holidays both at home and abroad, the industry's chief hope lies with foreign visitors who account for about a quarter of all the country's holiday makers. The national tourist board reports that 104 'tourist projects' are currently under way which will add 22 000 hotel rooms to the existing total of 300 000. The occupancy rate which had dropped below 50 per cent in 1991, reached 83 per cent in 1994.

A TOURISM STRATEGY

With the strong backing of the World Travel and Tourism Council, a major overhaul of Brazil's tourism industry is underway. Embrateur, the organisation responsible for tourist policy has identified three major objectives:

- improvement in basic infrastructure in the regions that are designated for tourism;
- the need to improve the quality of service, so as to become competitive in an international market;
- the need to invest in marketing and promotion to change Brazil's image abroad. Caio Luis de Carvalho, Embrateur's president says 'we want to show that we are an emerging country, a modern country, a country that has fantastic things. A country that has social problems, but which are the results of the past and which we are trying to change.'

Embrateur aims to double the number of foreign visitors to Brazil by 1999 (from 1996). That will entail increasing employment in the tourist sector by about a fifth. European tourists in particular are being targeted: for example only about 30 000 British tourists visit Brazil each year. At the same time, the stabilisation of the Brazilian currency and increasing purchasing power have led to a sudden growth in internal tourism in Brazil. The number of Brazilians travelling abroad has also risen markedly, which makes it all the more necessary to raise the number of incoming tourists.

Figure 8.2 Characteristics of Brazilians travelling abroad

(% UNLESS OTHERWISE INDICATED)	
Trip purpose	
Tourism	67.7
Business	23.3
Conference/convention	4.3
Other	4.7
Organisation of the trip	
Not organised by an agency	54.3
Organised by an agency	45.7
Accommodation used	
Hotel	74.0
House of friends/relatives	19.4
Rented apartment	2.8
Other	3.8
Tourists who in last 12 months	
Have travelled in Brazil	63.4
Have not travelled in Brazil	36.6
Average length of stay (days)	17.54
Average per head daily expenditure (US$)	80.02
Permanent residence of the tourists	
São Paulo	32.8
Rio de Janeiro	24.3
Rio Grande do Sul	11.6
Minas Gerais	5.4
Paraná	4.9
Pernambuco	4.5
Bahia	3.7
Federal district	2.0
Other	10.8
Professions	
Engineer	10.9
Businessman	9.6
Teacher	9.3
Administrator	5.4
Doctor	4.8
Lawyer	4.5
Business owner	4.4
Other	51.1

ECO-TOURISM

In recent years eco-tourism has become a not insignificant sector of the industry. The federal government has produced a national policy on eco-tourism which recognises both the potential and problems of opening up the country's ecologically exceptional areas. Some areas, notably those reserved for Indian communities, are strictly off-limits.

Figure 8.3 The main tourist attractions in the five regions

NORTHERN REGION

If you are looking for exotic sights, visit the north of Brazil. You will be overwhelmed by the natural beauty of this paradise on earth.

The architecture of Manaus, capital of the state of Amazonia, is highly impressive. The façade of its famous theater, the "Teatro Amazonas", is neo-classical, but inside the decor is art nouveau. The city's hotels and its restaurants with their typical dishes provide tourists with the necessary infrastructure for a pleasant stay. The big attraction, however, is spending one or two nights in the jungle to get the feel of it. But there is no need to rough it: hotels and guides are available, and the safety of tourists is assured.

Belém, the capital of the state of Pará, is a major trade and economic center.

Another great attraction is the Island of Marajó, famous for an unusual tidal phenomenon.

The Bananal Island in the state of Tocantins is the largest river island in the world, known for its lovely beaches and good fishing.

NORTHEAST REGION

The Northeast is rich in folklore and tradition.

Salvador, capital of the state of Bahia, is said to have 365 churches. The city's Carnival is an event of and for its people at which tourists are welcome to participate. It is one of the region's major attractions.

Due to the canals that abound in the city, Recife, the capital of the state of Pernambuco, is often called Brazil's Venice. Recife also has beautiful beaches.

Olinda, a historical town, has long ago become an extension of Recife. In 1982, UNESCO listed it as a Cultural and Natural Heritage of Humanity.

Fortaleza, capital of the state of Ceará, has super-modern buildings.

Next comes Maceió, the capital of the state of Alagoas. The city's beaches such as Jaiúca and Pajuçara, are considered to be the most beautiful in Brazil.

João Pessoa is the capital of the state of Paraiba. One of its major attractions is the Cabo Branco lighthouse.

In Teresina, the capital of the state of Piauí the main attraction is the National Park of Seven Cities.

CENTRAL WEST REGION

In this Region lies Brasilia, the capital of Brazil. The city is listed as a historical heritage of humanity. Due to its ultra-modern architecture, it is one of the most beautiful cities in the world. There is harmony among its buildings, such as the Alvorada Palace, the Metropolitan Cathedral and the National Congress.

Cuiabá, the capital of the state of Mato Grosso is the gateway to the Pantanal, the Wetlands. It, like the Amazon Jungle, is one of the world's largest reservations of fauna and flora. Rare species of birds and animals populate the region's countless rivers and lakes.

Goiânia, the capital of the state of Goiás, is one of Brazil's newest cities. It was built in 1933.

Once you are in the Caldas Novas spa, you can take it easy, lying mmersed in its famous curative waters.

SOUTHEAST REGION

The principal gateway to Brazil.

Rio de Janeiro is full of attractions: the Sugarloaf, the Christ Mountain, the Botanical Gardens, the Tijuca Forest, museums, churches and countless beaches: Copacabana, Ipanema, Barra da Tijuca.

The Rio Carnival, celebrated in February, is famous throughout the world for its Samba School parade.

In the south of the state, along the "Costa Verde" (Green Coast) are many sandy beaches. Crossing Guanabara Bay, you are on the road to the "Costa do Sol" (Sun Coast) and its resorts.

São Paulo is South America's major commercial, industrial and economic center. Over 600 hotels and close on 1000 restaurants are ready to cater to visitors.

Belo Horizonte is the capital of the state of Minas Gerais where Brazil's architectural past has been preserved. Of special interest are the sculptures of Alejadinho and the state's historical towns.

SOUTHERN REGION

The South is the only region in Brazil where a change in season is clearly defined.

The state of Paraná offers tourists an unforgettable sight: the world famous Iguassú Waterfall.

Florianópolis is known for its gorgeous beaches: "Canavieiras", "Ingleses", "Armação", and "Joaquina"

Porto Alegre is the capital of Rio Grande do Sul, a state typical for the south of Brazil. The Gauchos, as its people are called, provide the finest gastronomic delicacy of the region: the "churrasco" or barbecue.

Tramandaí, Capão da Canda, Atlântida and Torres are some of Rio Grande do Sul's most beautiful beaches.

The hill towns of this state are well worth visiting: Caxias do Sul, Farroupilha, Bento Gonçalves, Canela and Gramado.

However in other locations the government view is that development can achieve both socio-economic and ecological goals. Bringing in small numbers of ecologically-minded tourists can help diversify the economies of poor, sparsely populated areas. By encouraging the establishment of small businesses and otherwise creating employment, local people are more likely to remain living where they are.

? ? ? ? ? ? ? QUESTIONS ? ? ? ? ? ? ?

1. Suggest why there was little change in receipts on Brazil's international travel account between 1992 and 1994 while expenditure increased by more than 50 per cent (Figure 8.1)
2. Analyse Figure 8.2 which shows the characteristics of Brazilians travelling abroad.
3. Outline the essential contrasts between the five regions in terms of tourist attraction.

Rio de Janeiro – the battle between positive and negative images

SETTING

Rio de Janeiro means 'River of January', commemorating 1st January 1502 when Andre Goncalves, a Portuguese captain sailed into Guanabara Bay (Figure 8.4), thinking he was heading into the mouth of a great river. The people of Rio often refer to their city, the country's number one tourist destination, as the 'Cidade Marvilhosa'. This is mainly because of a warm climate with sunshine during most of the year and an extensive coastline with numerous beaches. The urban area has been moulded around the foothills of the mountain range which provides its backdrop. In the bay there are many rocky islands fringed with white sand which add significantly to the panoramic view of Rio which is usually described as breathtaking.

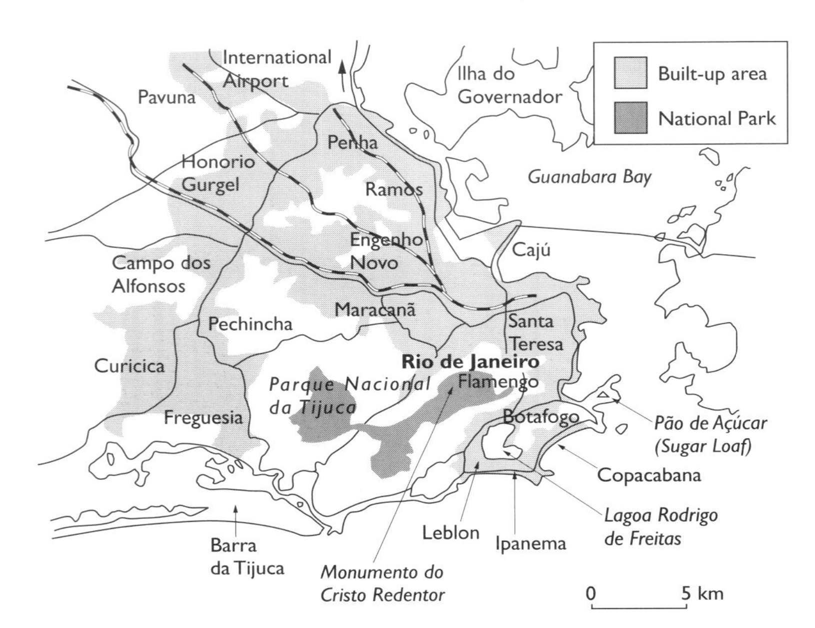

Figure 8.4 Map of Rio de Janeiro

Rio combines world famous landmarks with the international perception of great style. The city first gained global renown as a leisure destination in the 1930s, when it received the patronage of the first-generation jet set, subsequently figuring prominently in Hollywood films. The City of Rio de Janeiro, capital of the State of Rio de Janeiro, has 5.5 million inhabitants, about 40 per cent of the state's population. The Metropolitan Region of Rio de Janeiro has a population of 9.6 million. The second largest urban area in the country, after São Paulo, Rio was replaced by Brasília as the national capital in 1960.

Figure 8.5 Panoramic view over Rio showing Christ the Redeemer and the Sugar Loaf

CLIMATE

The most popular time to visit in terms of climate is between May and August when the region is cooled by trade winds and the maximum daily temperature is 24–25°C (Figure 8.6). Between December and March, the rainy season, the climate is much more humid with accompanying higher temperatures. However, even in the latter period it's never as oppressive as it is in the north of Brazil.

Figure 8.6 Weather data for Rio de Janeiro and London

Rio de Janeiro Brazil 60m		J	F	M	A	M	J	J	A	S	O	N	D	YEAR
Temperature	Daily Max. °C	29	30	29	27	25	25	24	25	24	25	26	28	26
	Daily Min. °C	23	23	22	21	19	18	17	18	19	19	20	22	20
	Average Monthly °C	26	26	25	24	22	21	21	21	21	22	23	25	23
Rainfall	Monthly Total mm	125	122	130	107	79	53	41	43	66	79	104	137	1086
	No. of Days	13	11	12	10	10	7	7	7	11	13	13	14	128
Sunshine	Hours per day	6.9	6.9	6.8	6.3	6.2	6.3	6.5	6.6	5.1	5.1	5.7	5.6	6.2

London (Kew) United Kingdom 5m														
Temperature	Daily Max. °C	6	7	10	13	17	20	22	21	19	14	10	7	14
	Daily Min. °C	2	2	3	6	8	12	14	13	11	8	5	4	7
	Average Monthly °C	4	5	7	9	12	16	18	17	15	11	8	5	11
Rainfall	Monthly Total mm	54	40	37	37	46	45	57	59	49	57	64	48	593
	No. of Days	15	13	11	12	12	11	12	11	13	13	15	15	153
Sunshine	Hours per day	1.7	2.3	3.5	5.7	6.7	7	6.6	6	5	3.3	1.9	1.4	4.3

COMMUNICATIONS AND TOURIST INFRASTRUCTURE

Galeao, Rio de Janeiro's international airport, is the most important and busiest in Brazil. About five million passengers passed through Galeao in 1992. The Santos Dumont Airport in the heart of the City, handling domestic flights only, ranks fourth among Brazilian airports. An extensive highway infrastructure gives access to all parts of the city and there is a diversified public transport system which consists of approximately 6000 urban buses, 250 km of urban railway, and 24 km of subways, aside from water transportation in Guanabara Bay.

Rio advertises itself as Brazil's main gateway for foreign tourists. It hosts annually about two million visitors who benefit from its extensive tourist infrastructure and the relaxed lifestyle which the city is keen to project (Figure 8.7). Rio is among the world's cities that hosts the greatest number of trade and scientific events – fairs, symposiums, congresses and exhibitions. The city has an extensive network of theatres, night clubs, cinemas, museums, libraries and art galleries. It also hosts Carnival, which, according to the city's own publicity 'is the synthesis of the cultural expression of a people that discovered the secret of how to cheerfully enjoy work and play and to share its experiences with visitors'. Carnival in Rio, which occurs on the eve of Lent, is big business. Large firms hire VIP boxes for guests who are flown in from all parts of Brazil and abroad.

A significant urban reconstruction programme is in progress which will enhance the city's facilities as a whole, particularly along the main commercial corridor. Drainage and sewage systems have been overhauled and there has been a lot of work on roads. Several new tunnels have been bored through the hills to improve general traffic movement. The economic stability of recent years has spurred retail investment. For example Rio is currently in a shopping mall 'boom' (Figure 8.9).

Rio sets the fashion, the habits and way of life in Brazil

The way to start the day in Rio de Janeiro is to get out of bed a little after dawn as the mist circles the middle reaches of Sugar Loaf, bounce through the hotel lobby clad only in your scantiest swimwear and, after speedwalking in hip-wiggling style to the beach nearby, plunge into the Ipanema surf. There's no need to be self-conscious, just as long as what you are wearing is in line with this season's style and colour and, crucially, as brief as possible. In Rio, they mark a man by the cut of his tanga.

For more than 30 years Rio has developed a lifestyle which gives a general impression of frivolity – for the moneyed classes at least. Little wonder, given that it lost its status as Brazil's political capital in 1960 to Brasilia, has ceded power to São Paulo as the economic and business capital and is now having its lead eroded as tourism capital by the pristine beaches of the Brazilian north-east. Oblivious, Rio's citizens, or *Cariocas* as they're known, blithely do the samba till all hours of the night, smiling endlessly.

The last 20 to 30 years, however, have been marked by sharp cuts in both state and federal funding to Rio. Thus the slums (*favelas*) on almost every hillside cling miserably to life and the city's infrastructure struggles to survive. The clean and efficient metro is a case in point. Begun in 1975 and never completed, it fails to extend to such obvious destinations as Copacabana and Ipanema beaches.

For tourists and the better-off *Cariocas* this is probably a blessing; easier access by metro would surely mean even more *favela* dwellers hassling them on the beaches. Downtown Rio, the business centre by day, has already been largely abandoned by the middle and upper middle classes after dusk.

Foreign residents and chambers of commerce moan they are "sick and tired" of journalists who give Rio a bad name. There has been very little violent crime directed at tourists and business travellers, and as the expats trumpet, "We are no Miami". But the fear of petty theft downtown and by the beaches is real. People apt to complain of exaggerated stories from abroad one minute, will strongly insist you hang firmly on to your briefcase the next.

Which is, of course, all a great shame.

Rio is, along with Mexico City and Buenos Aires, one of Latin America's three truly world-class cities and is the most beautiful. Take the cable car up Sugar Loaf for the supreme view (superior even to that from the Christ the Redeemer monument on Corcovado) of how Rio tucks itself into the limited space between sea and mountain and along a chain of inlets and bays.

The beaches improve as you head south-west away from downtown. The huge crescent of Copacabana is the most stunning, though the Japanese project to clean up the surrounding bay has been hatched not a moment too soon. A dead furry creature floating by my knees in Copacabana's diesel-gummed waters would surely have agreed.

However, after several seasons when Rio's image problems caused visitors to stay away, things are looking up. The hotels reported that the February carnival (traditionally a sell-out by the previous September) was the first in the last three years to have filled the hotels again.

Business Traveller, May 1994

Figure 8.7 Rio – big fun and petty crime

Figure 8.8 Carnival scene

Rio builds 14 shopping malls

The Rio de Janeiro shopping mall market is currently in an expansion phase. Construction projects reaching to the year 1998 total some R$ 710 million. Fourteen malls are to be constructed.

Emílio Habib, director of Tenant Mix, a company that administrates shopping malls in Rio de Janeiro, said that there were 12 other shopping malls that have not been announced in addition to the fourteen that will be inaugurated over the next few years.

According to Habib, the majority of the undertakings are expected to be concluded by 1997 and practically all of them have their store space sold. Habib said that 10 shopping malls have been inaugurated in Rio de Janeiro over the last two years and that the Real Plan was responsible for growth in the sector.

Gazeta Mercantil, 26 August 1996

Figure 8.9 Shopping mall boom in Rio

FUNCTIONAL ZONES

The city is divided into three parts:

1. Centro, situated on the bay, is the historical and commercial centre of Rio. The centre of the city has been rebuilt many times since colonial days but although the elegance of its colonial and neoclassical architecture has been overshadowed by modern, towering office buildings, it has by no means been completly swamped.
2. Zona Norte contains the city's industrial areas and the major working-class residential 'bairros' (neighbourhoods), with little to offer in terms of historical interest or natural beauty. The Rio-Niteroi Bridge, about 14 km long, connects the zona norte with Niteroi, a city east of the bay.
3. Zona Sul is the name used to cover the area south of the city centre, though it is generally taken to mean just the bairros shouldering the coastline. It is here that the tourist industry is concentrated. For more than 60 years the beaches have been Rio's heart and soul, providing a constant source of income and recreation for the 'cariocas', the inhabitants of the city. The value that Brazilians attribute to their beaches should not be underestimated. Brazil has one of the longest continuous coastlines in the world, 7700 km, much of it comprising sandy beach. These beaches have made an important contribution to a sense of 'easy living' and pleasure even for many of the poor. Some sociologists have viewed the beaches as an important political safety valve.

GEOMORPHOLOGICAL HIGHLIGHTS

The Corcovado ('hunchback') mountain is undoubtedly the symbol of Rio. On top of its 710 m is the famous statue of Christ the Redeemer with its arms outstretched in welcome. The statue, finished in 1831, weighs over 1000 tonnes and is 30 m high. The Corcovado is a part of the 3300 hectares of Tijuca National Park located right in the middle of the city. Of almost equal renown is the 394 m Sugar Loaf mountain, located at the mouth of Guanabara Bay, which can be reached by cable car.

THE BEACHES

Figure 8.10 Beach scene illustrating the Tanga

However it is the beaches where visitors spend most of their time. The city's beaches begin in Guanabara Bay and stretch out to the Atlantic Ocean in an uninterrupted sequence. First come Flamengo and Botafogo, followed by Urca and Praia Vermelha. Leme Beach stretches into the glistening sands of Copacabana, Brazil's most famous beach, but less classy than it once was, losing ground to its southern neighbours. Avenida Atlantica runs along the beach with bars, restaurants and sidewalk cafes. The 4.5 km of sand end at Fort Copacabana. South of the fort are Diabo Beach and Arpoador Rock. Here Ipanema Beach begins. It was in Ipanema that the 'tanga', a diminutive version of the bikini first appeared on the international fashion scene. Leblon Beach follows Ipanema, ending 2 km later at the foot of Dois Irmaos Rock. Further along the coastline is São Conrado Beach. Afterwards, Barra Beaches offers 17 km of beautiful, uncrowded sands in an area that is quickly becoming the new attraction in Rio. Beyond Barra are even emptier beaches like Recreio dos Bandeirantes, Prainha, Grumari and Guaratiba.

On the other side of Guanabara Bay is the city of Niteroi. It is the starting point to the secluded beaches of the Sun Coast. The region has many attractive beachfront resorts such as Marica, Saquarema and Araruama.

PROTECTED AREAS

In order to control urban expansion and protect areas of ecological importance, various state departments control 29 environmental preservation areas, 17 permanent preservation areas, three reservations, 19 historical monuments, 11 parks and 25 protective forest areas. Above all, Rio boasts Tijuca Forest, claimed to be the biggest urban forest in the world.

NEGATIVE IMAGES

Set against Rio's array of attractions are a number of significant negative images which have undoubtedly had considerable repercussions on the city's tourist industry. The most serious of these are clearly social but there are also a range of environmental problems associated with continued urban growth. There is of course an overlap between the two sets of factors but the essential issue is that both have an adverse impact on the outsider's perception of the city:

- Urbanisation continues at a rapid rate, pushing the city along the coast in both directions. The dualism exhibited by this urban environment has become more and more obvious to visitors (Figure 8.11). In common with most mega-cities in the developing world, a major element of Rio's urban problems has been the irregular occupation of land, of which the favelas are the most visible example. The city authorities have recently pursued a twin-track policy aimed at improving the situation. While shanty dwellings in areas of environmental importance are being removed, informal communities in less sensitive areas are being integrated into the city (Figure 8.12). Such areas are being recognised as valid neighbourhoods and proper urban services are being provided.
- Rio's rich architectural heritage is still being whittled away with the demand for more commercial space in and around the central area.
- A growing population, a wide range of industry, and increasing levels of traffic are all having a detrimental effect in terms of land, sea and air pollution. If reality moves too far away from the image of a sub-tropical paradise, then the number of visitors will certainly decline.
- In recent years Rio has acquired the image of a dangerous place (Figures 8.13 and 8.14) with homicide rates rivalling those of the most troubled US cities. Although major acts of violence are rarely perpetrated on visitors, the guide books spell out the risks and the areas where the visitor should be particularly wary.

Figure 8.11 Tourism and voyeurism

When in Rio head for the slums

VISITORS to Rio de Janeiro have traditionally headed for Copacabana Beach, the giant statue of Christ on Corcovado and February's exotic carnival. But an increasing number are adding a new destination: the city's shantytowns.

Until recently, any tourist visiting a *favela* — as the shantytowns are known — would almost certainly have been mugged. Lone visitors would still be at risk, but tours with a local guide are now being organised.

There are more than 600 *favelas* around Rio. They have been part of the city since the start of the century, when federal troops, after putting down a rebellion in the north-east, were discharged and came to the city, setting up shacks on a hillside near town. The favelas have no running water or paved roads and today are ruled by heavily armed drug dealers, but schools, medical centres and churches have sprung up within them.

The largest in Brazil is Rocinha, home to more than 300,000 people, who live in shacks clinging to a mountain behind one of Rio's luxury hotels, the Inter-Continental on São Conrado beach. I asked the concierge at my hotel, the Copacabana Palace, to arrange an excursion, and within an hour was setting off with a guide.

My guide drove me up the steep mountain. With each curve of the dirt track the smell of rotting rubbish increased. My guide assured me that as everybody knew him I was safe.

He was right. We were barely given a second glance.

We walked up steep alleyways between homes of corrugated iron, wood and rubble, and visited a school and a clinic. I had assumed the locals were getting a cut of the £30 fee for the two-hour tour – but they were not. The guide claimed he brings a kilogram of flour or sugar to a family now and again in return, and said the locals liked outsiders to see how well organised and dignified their world was.

But I felt embarrassed because we were disrupting classes and peering into homes without giving anything in return.

Sue Wheat of the pressure group Tourism Concern points out that such experiences can be avoided by doing some research and choosing a tour in which local people are involved.

Paulo Veloso of the tour operator Journey Latin America said it would only consider arranging visits to *favelas* for special groups or individuals, such as doctors, who could bring knowledge or skills to benefit the locals. People should not be taken swiftly around a favela as if it were a ''human zoo''. ''You need to look at the *favelas* in greater depth than a day trip allows.''

Figure 8.12 Government plans to improve favela quality of life

The favelas that blanket many of Rio's hillsides have gained notoriety as places of extreme poverty and hopelessness. But a new project hopes to change all this

Until recently, the favelas were commonly regarded as no go areas. But the city government has now set up the $300 million Favela Bairro project to improve living conditions for their inhabitants. Over the next few years, 66 of the cities' 560 favelas will be integrated into the city. Initially, 16 mid sized slums – not those with the biggest drug and unemployment problems – have been selected by Sergio Magalhães, municipal secretary for housing, and his team. The biggest is Complexo do Andaraí, near Flamengo, where almost 7,000 people live in 1,740 houses. They will widen streets, put in pavements, lay pipes for water and cables for electricity, improve sanitation and build squares and football pitches. This is a new approach; in the past, favelas were often simply bulldozed. Today, people are only moved from those homes at risk of subsidence or those considered to be a health hazard. New homes are built for them in the same community if possible. Each of the slums has different problems associated with its physical geography – landslides in favelas built on the steep hillsides and drainage problems in riverside slums are particular concerns.

''What we want to do is to transform the favelas, not destroy them but make them part of the city both socially and culturally,'' says Magalhães. ''The programme is one of integration. Each favela has its own style of architecture and we want it to retain its own look but also to have all the services necessary to modern life.'' In return, the residents will have to pay taxes.

In Caju, a peninsula on the northern shore of Guanabara Bay, is the São Sebastião Park and Ladeira dos Funcionarios favela, home to about 800 families (4,000 people). Some of them have lived here for more than 40 years, when they first came to work at the nearby São Sebastião hospital. At the top of a steep hill, is the construction company that is gradually transforming the area into a liveable part of the city. Half of the workers are from the favela itself, learning the trade so that they can find jobs in building. Old wooden houses and those that are slowly slipping away are being torn down and replaced with sturdy new ones. The houses will measure 3.5 metres by 6 metres and include a yard of the same size – spacious in favela terms.

Figure 8.13 The divide of Rio

OVER the past six years, according to a throw-away paragraph in the normally-accurate *Jornal do Brasil*, more people have died by violence in Rio de Janeiro's sprawling suburbs than the 58,000 American fatalities in Vietnam.

True of false, there is no need to cancel your sun-soaked Copacabana holiday. The vast bulk of the violence happened in the ''other Brazil'' — the Brazil of *favella* shanty-towns, largely shielded from visitors' eyes by newly-built fly-overs on the way in from Galleao International airport.

Down in the palm encrusted, golden-beached South Zone, the limousines are still polished, the maids and porters are still deferential and, as long as you leave your Nikon and Rolex at home, you will enjoy the lotus-eating holiday of a lifetime.

It was Mario Henrique Simonsen, Citibank director and the country's best-known economist, who christened Brazil ''Belgindia'' — a combination of two parallel nations, prosperous Belgium and sprawling, poverty-stricken India, that just happen to occupy the same territorial space.

Figure 8.14 A dangerous place

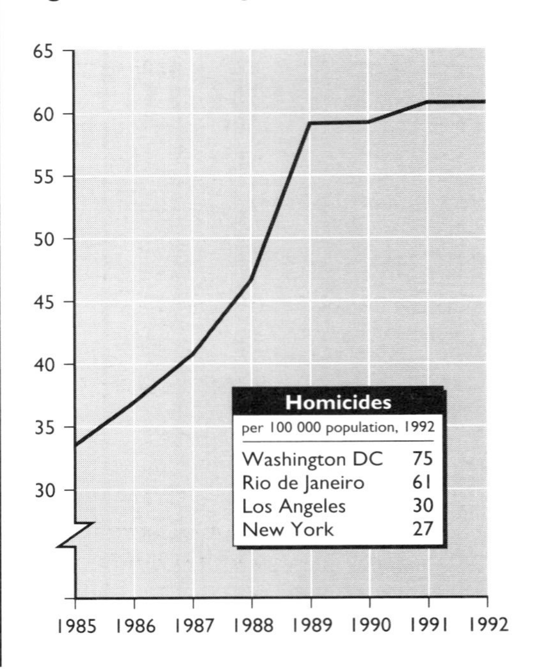

THE FUTURE

The importance of this sector of the economy is emphasised in a recent city publication which states 'Rio de Janeiro's future is tied to its consolidation as a service, trade, business, technology, education, tourism, leisure and cultural centre.' With the global expansion of the tertiary sector, few areas are growing faster than tourism. Rio is anxious to at least attract its fair share of such growth and to maintain its position as the country's premier tourist destination. The city has now set its sights on being the first in South America to host the Olympic Games; a site has been selected on Fundao Island, which is only a few minutes from the airport, and which could be ready early in the next century.

? ? ? ? ? ? ? QUESTIONS ? ? ? ? ? ? ?

1. Describe the location of Rio de Janeiro.
2. Analyse the climatic data provided. To what extent does climate make Rio a seasonal destination?
3. Detail the elements of the 'tourism infrastructure' you would expect to find in a major international destination such as Rio.
4. To what extent is Rio's tourist industry spatially concentrated in terms of its functional zones?
5. Suggest why some sociologists regard Brazil's beaches as an important safety valve.
6. (a) Discuss the negative images that Brazil's tourist industry has to contend with.

   (b) What reasonable policies might Rio pursue to reduce its perceived problems so as to reduce its areas of poor image?
7. Why would a successful bid for the Olympic Games be of long-lasting benefit to tourism in Rio?

Expanding Tourism in the Northeast

NEW INVESTMENT

The Northeast has gained more than any other region from the government's recent push to expand tourism. The National Economic and Social Development Bank has launched a $3 billion plan to boost the region, including subsidised financing for hotels. A separate project, backed by the Inter-American Development Bank, will put $800 million into infrastructure, ranging from sewerage systems to airport development. Both SUDENE and the private sector are keen to develop the potential of what could possibly become the premier international tourist destination in South America.

FIRST CLASS BEACHES AND COLONIAL HERITAGE

The coastline is over 2000 km of practically unbroken beach (Figure 8.24), much of it just what most people imagine tropical beaches to be – palm trees, white sands and warm, blue sea. The sunshine is virtually guaranteed. Visitors are also attracted by the region's colonial heritage which survives in the Baroque churches and cobbled streets of Salvador, Olinda and São Luis, often side by side with the modern Brazilian mix of skyscrapers and shanty towns. In Salvador and Recife, the Northeast has two of Brazil's great cities. Salvador, the capital of Bahia was, until 1763, also the capital of Brazil.

It is said that Salvador has 365 churches, one for each day of the year. The carnival in Salvador is a major attraction for tourists – a fancy dress revelry that lasts four days without interruption. Recife is known as the 'Brazilian Venice' because of its many canals and waterways and the innumerable bridges that span them. The name 'Recife' comes from the barrier reef ('*arrecife*' in Portuguese) that protects the city's beaches. Recife is the major gateway to the Northeast with regular flights to all major cities in Brazil as well as to international destinations, particularly Lisbon, London, Frankfurt and Paris.

Figure 8.15
Brazil's beaches, for the tourists and the locals

Where the beach is Brazil

Whether a passing tourist or local worker, Brazil's beaches are the place to hang out

WITH NEARLY 8,000 kilometres of extremely varied coastline, Brazil boasts many of the best beaches in Latin America. There's a beach to suit every taste. If it's people-watching you are into, then Copacabana in Rio still draws the crowds and the ratio of beautiful bodies per square metre beats even California hands down. Body-builders strut, while girls of every conceivable racial combination bare all but a token symbol of modesty in mini-bikinis aptly named 'dental floss'.

But if it is the celebrity that you are after, then you need to get away from the big cities, to the expensive up-market resorts like Buzios in Rio de Janeiro state, which was 'discovered' by Brigitte Bardot thirty years ago. It's the nearest you will get to the chic bohemianism of the French Cote d'Azur – but with a far bigger hinterland.

The jet-set may still favour the better-known resorts, many of which get unbearably busy during the high-season (December to February), but the more adventurous beach-bums have been pushing ever further north, through Bahia, with its endless succession of palm-fringed, deserted shores, up into Natal and Ceara, in search of the perfect place to get away from it all.

Over the past two decades, a whole series of former fishing villages in Ceara, in particular, have become 'paradise' for a few seasons. For a while, Canoa Quebrada, cut off by dunes and cliffs from the nearby town of Aracati, was the haunt of all that was hip. The local fishermen were welcoming; the food was fresh, if basic. And all one had to do for lodgings was to hang up one's hammock. But in the new communications age, the secret was soon out, and a situation developed that has become the pattern for so many of the formerly isolated communities along the North East coast.

Roads were built, buses arrived. The villages were put on tourist itineraries. The beach groupies turned their back on Canoa Quebrada when the 'respectable' tourists arrived, and moved on to Jericoquara – even more remote, even more unspoilt – for a brief while. But now that too is firmly on the tourist map. The local men, who used to take their little boats out to fish, are getting more adept at running simple guest houses or manning the ubiquitous little drinks stalls on wheels that seem to sprout on every beach in Brazil whenever any swimmer, foreign or local, has arrived.

Purists deplore the loss of the centuries-old simplicity among the coastal peoples. But, for many of the locals concerned, change has brought a prosperity that their forefathers could never have dreamed of. And, by European standards, it is still picturesquely primitive and very tropical. The Brazilian tourist authorities know the beaches are their strongest selling point.

The vigour of beach-life in Brazil is not just a phenomenon put on for the tourists, however. Unlike many other long-distance resorts such as Goa or Penang, against which Brazilian beach centres might wish to compete, Brazil has its own indigenous beach culture. At weekends, in particular, local people flock to the sea, often in family groups, loud music blaring from sound systems wired up to their car batteries.

Others make their own music and sing and dance, knocking back the rum, or rougher, cheaper equivalents, while hawkers mingle among the bronzing bodies, selling snacks. Impromptu football or handball games break out; from time to time, the roar of voices and the sea is broken by the buzz of a beach-buggy racing across the sand. For tens of millions of people living in the coastal cities and villages, the beach is Brazil.

Brazil, [Times supplement], 9 December 1996

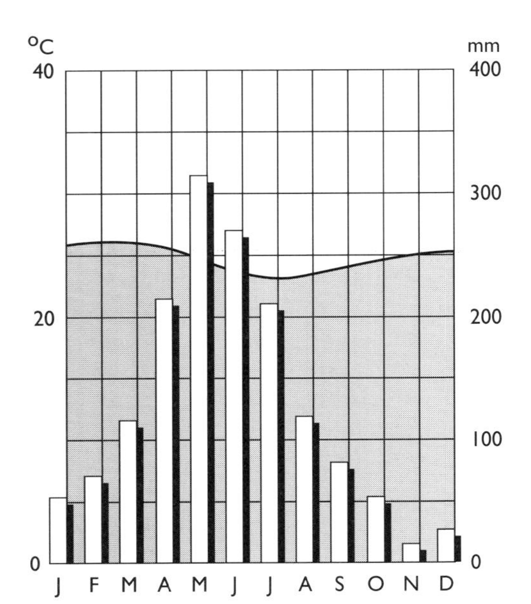

Figure 8.16 Climate graph for Maceió

THE INTERIOR

Inland of the flat coastal strip, the 'zona da mata', is an intermediate area, the 'agreste', where hills rear up into rocky mountain ranges and the lush, tropical vegetation of the coast is gradually replaced by highland scrub and cactus. Then comes the 'sertão', the vast semi-arid interior that covers more than three-quarters of the Northeast. In the sertão, Portuguese and Indian influences predominate while in the coastal area African influences become very obvious. The colonisation of Brazil began in the Northeast, ensuring that the region is rich in folklore and tradition. However, to date the bulk of the investment in tourism has gone to the coast. Not until the industry 'matures' will the interior command more attention.

Figure 8.17 Colonial architecture, the Justice Building, Recife

Thus at present the interior is the preserve of the more adventurous, personally organised tourist (Figure 8.27).

MARKETING STRATEGIES

Marketing has played an important role in the development of tourism. For example:

- visitors can take out a 'sun insurance' policy, guaranteeing an extra day with no charge for every rainy day during their stay;
- samba dancers in Rio's world famous carnival were recently paid to sing about the positive environment for tourism in Ceará;
- a former governor of Ceará state has been the major backer of a TV soap opera filmed on Ceará's beaches.

The major players in the industry are only too well aware of how intense competition is in the tourist market both nationally and internationally.

INTERNATIONAL HOTEL GROUPS MOVE IN

The Northeast received one million visitors in 1994, who spent $497 million, a tenfold increase from 1990. The region's total receipts from tourism doubled to $1 billion between 1991 and 1994. Beach Park, an $18 million aquatic park, handled 1.2 million visitors and $22 million in sales in 1995, almost three times the 1994 figures. International hotel groups such as Sheraton, Ramada, Westin, and Best Western have ventured into the region, but generally to manage rather than build hotels. If this 'testing the water' exercise proves successful there is little doubt that these big players will develop in a significant manner. However the high interest rates required to protect the economy against the traditional problem of inflation makes hotel construction expensive and results in high prices for foreign visitors. But if the government's economic reforms are successful, interest rates should fall, attracting more foreign visitors and Brazil's own increasingly rich middle-class.

WATER PARKS

Apart from hotels, most other elements of tourist infrastructure are also planned. For example, the Canadian Company Whitewater West Industries has formed its first joint venture with Brazilian Transversal Marketing to plan five new water parks in Brazil, two in the Northeast. The newly formed company, White Water Brazil is working on projects in Recife, Salvador, Gioânia, Rio de Janeiro and Brasília. Recife's Acquamundi project will require a total investment of $25 million but should begin repaying debts within three to five years. It has been estimated that for every person employed within the water park, five indirect jobs will also be created outside the park.

THE GOLDEN COAST PROJECT

With over 60 per cent of tourists to the Northeast demanding beach holidays, the state of Pernambuco is well placed to increase its tourist revenue. A project of considerable significance is the 'Costa Dourada' (Golden Coast), located 65 km south from Recife's international airport (Figure 8.19). Here lie the best beaches in the state. Jesus Camara Zapata, Pernambuco's director of special projects is convinced that this development will soon be on the European tourist trail.

The Golden Coast, with an average sea temperature of 28°C, provides more than 15 km of coastline, alternating wide open sea with sheltered reef beaches, coconut groves, mangroves and forests. It has been based on the best aspects of international tourism in Cancun (Mexico), the Dominican Republic and Polynesia.

Figure 8.18
Off the beaten track tourism

Into the sertão

The Pernambucan *sertão* is one of the harshest in the Northeast, a scorched landscape under relentless sun for most of the year. This is cattle country, home of the *vaqueiro*, the North-eastern cowboy, and has been since the very beginning of Portuguese penetration inland in the seventeenth century: one of the oldest frontiers in the Americas.

Travelling in the sertão requires some preparation, as the interior is not geared to tourism. Hotels are fewer and dirtier; buses are less frequent, and you often have to rely on country services which leave very early in the morning and seem to stop every few hundred yards. A **hammock** is essential, as it's the coolest and most comfortable way to sleep, much better than the grimy beds in inland hotels, all of which have hammock hooks set into the walls as a standard fitting. The towns are much smaller than on the coast, and in most places there's little to do in the evening, as the population turns in early to be up for work at dawn. Far more people carry arms than on the coast, but in fact the *sertão* is one of the safest areas of Brazil for travellers – the guns are mainly used on animals, especially small birds, which are massacred on an enormous scale. Avoid tap water, by sticking to mineral water or soft drinks: dysentery is common, and although not dangerous these days it's extremely unpleasant.

But don't let these considerations put you off. People in the *sertão* are intrigued by gringos and are invariably very friendly. And while few *sertão* towns may have much to offer in terms of excitement or entertainment, the landscape in which they are set is spectacular. The Pernambucan *sertão* is hilly and the main highway, which runs through it like a spinal column, winds through scenery unlike any you'll have seen before – an apparently endless expanse of cactus and scrub so thick in places that cowboys have to wear leather armour to protect themselves. If you travel in the rainy season here – March to June, although rain can never be relied upon in the interior – you may be lucky enough to catch it bursting with green, punctuated by the whites, reds and purples of flowering trees and cacti. Massive electrical storms are common at this time of year, and at night the horizon can flicker with sheet lightning for hours at a stretch.

Brazil: The Rough Guide

The project has strict development rules, being determined to avoid the mistakes of mass tourism, exemplified perhaps at their worst in Spain and Florida. For example:

- no more than 2 per cent of the 10 000 hectares can be built on;
- there is a two-storey maximum building height and construction has to be in a traditional style;
- the environmental considerations extend to a Mangrove Research Centre sited along the estuary.

Apart from illustrating the beautiful and varied environment, marketing will stress:

- that Brazil is closer to Europe than Cancun (Mexico) or California;
- that the Northeast isn't a one-stop destination as many Caribbean islands are;
- for tourists from southern Brazil, Argentina, and Chile, this is the closest tropical destination.

However tourists will be shielded from the deprivation that is clearly evident in the nearby towns of Tamandare, Rio Formosa, Sirinhaem and Barra do Sirinhaem. The entrances to the beaches will have security controls and like many Caribbean resorts, visitors will not be encouraged to leave the reserve. While the neighbouring areas will not benefit directly from profits generated by the project, jobs will be created and there will be considerable improvements in roads, sewage and water supply.

The first hotels opened in mid-1997. The federal and state governments who have provided half of the $70 million infrastructure costs estimate that the project will attract a million visitors a year, generating more than $500 million in receipts. If successful, the Golden Coast project will provide a model for many more developments in the region.

Figure 8.19 The Golden Coast project

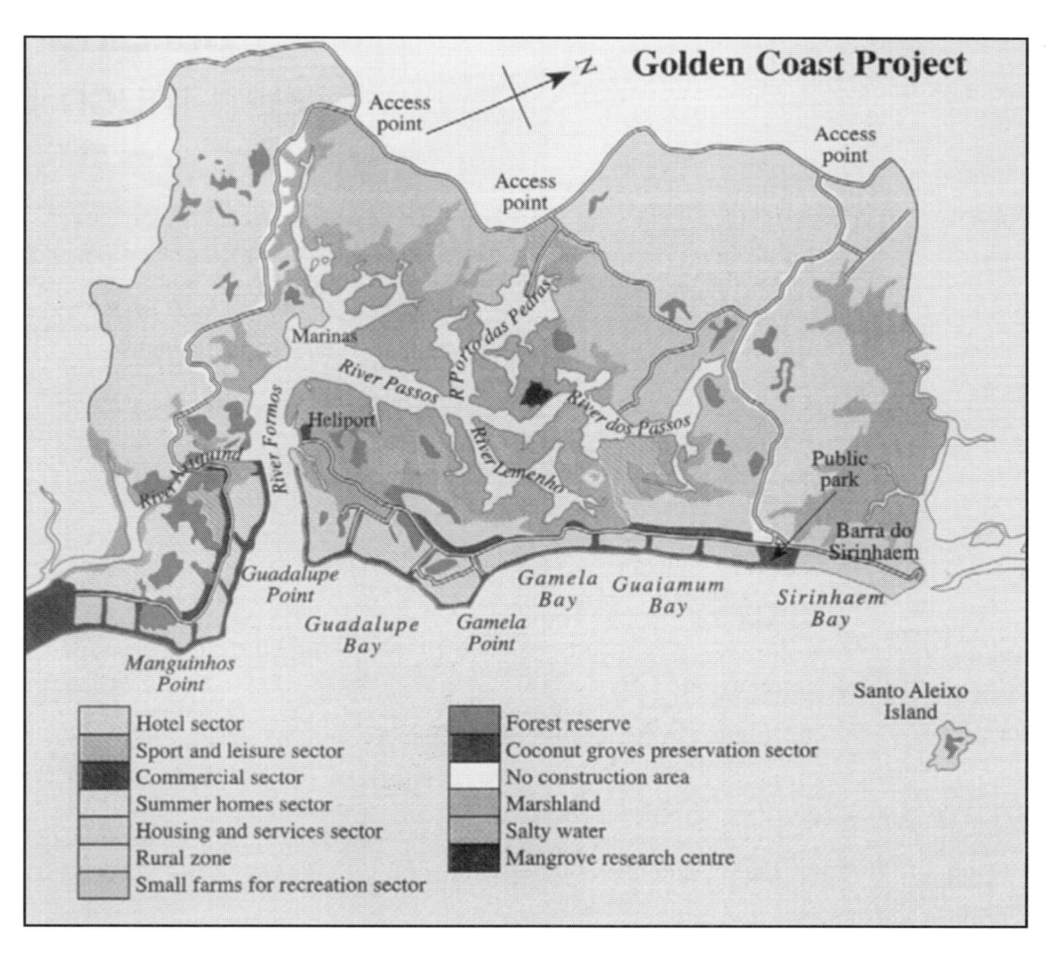

Figure 8.20 The Golden Coast project

QUESTIONS

1. Why has the Brazilian government identified the Northeast as the region with the greatest potential for tourist development?
2. To what extent and why does the climate of Maceió (Figure 8.16) differ from the climate of Rio de Janeiro?
3. How significant is the decision to build two large waterparks in the region?
4. Assess the scale and significance of the Golden Coast project.
5. What else does the Northeast have to offer the tourist apart from beach holidays?
6. What are the disadvantages of expanding the tourist industry in the Northeast?

Index